Linear Integrated Circuit Applications Manual

Linear Integrated Circuit Applications Manual

Robert J. Traister

ACADEMIC PRESS, INC.
Harcourt Brace Jovanovich, Publishers
San Diego New York Berkeley Boston London Sydney Tokyo Toronto

National's products are not authorized for use as critical components in life support devices or systems without the express written approval of the president of National Semiconductor Corporation. As used herein:

1. Life support devices or systems are devices or systems which, (a) are intended for surgical implant into the body, or (b) support or sustain life, and whose failure to perform, when properly used in accordance with instructions for use provided in the labeling, can be reasonably expected to result in a significant injury to the user.
2. A critical component is any component of a life support device or system whose failure to perform can be reasonably expected to cause the failure of the life support device or system, or to affect its safety or effectiveness.

Academic Press, Inc.
San Diego, California 92101

United Kingdom Edition published by
Academic Press Limited
24–28 Oval Road, London NW1 7DX

Library of Congress Cataloging-in-Publication Data

Traister, Robert J.
Linear integrated circuit applications manual / Robert J. Traister.
p. cm.
Includes index.
ISBN 0-12-697403-9
1. Linear integrated circuits. 2. Electronic circuit design.
I. Title.
TK7874.T74 1989
621.381'5—dc20 89-6844
CIP

Printed in the United States of America
89 90 91 92 9 8 7 6 5 4 3 2 1

This book is dedicated to the memory of Edith Colbert.

Contents

Preface

It has been said that the decade of the 1980s will be known as the digital decade. For it was during this 10-year period that the microcomputer became not only a household word but a household device. Exciting new digital technology advances were announced. Indeed, the digital age was upon us, starting in the previous decade with the coming of the digital watch, the digital calculator, and even digital audio taping techniques.

With all this talk and hoopla about digital-this and digital-that, the somewhat older linear technologies seemed to fall by the wayside. What seemed to be and what was and is, however, are two diverse pictures. Today linear IC technology and devices are indispensable when dealing with nearly every branch of electronics design.

The linear IC has endeared itself to every segment of society. Even today's "wholly" digital devices are powered by circuits that use linear IC voltage regulators as their prime regulating components. Nor has linear IC technology stood still while digital technology was getting all the applause. Today the idea of an AM or FM radio on a single chip is not out of the question. Rather, it can be ordered out of a catalog. The television receiver is becoming a more compact and less complex device, because linear IC building blocks are simply soldered in place instead of several hundred discrete components.

These advances have strengthened two very important prerequisites for successful electronic gadgetry. First, design, construction, and testing problems are minimized. This means that fewer man-hours are required and designs can be marketed more inexpensively. Second, fewer discrete components mean fewer wiring interconnections which, in turn, mean superior circuit performance and reliability. The end result is "cheaper and better," a combination that usually guarantees success.

And what versatility! Voltage regulators, operational amplifiers, buffers, voltage comparators, instrumentation amplifiers, video amplifiers, switches, filters, multiplexers, temperature sensors, sample and hold, voltage references, and many, many more fall into the linear IC category. This yields radio-frequency circuits, video circuits, audio circuits, motion-control circuits, and circuits that can and will do anything that has been done before and do it better; and things that were deemed impossible or highly unlikely a decade or so ago are now not only possible but practical.

This book has been written to serve a very direct and simple purpose. It is a sort of "shoppers' guide" to practical (in most cases) linear IC designs. Having served as a "benchman" and consultant, I know of the need for a convenient way to lay hands on a required circuit design. This book probably has it. Are you looking for a simple switching power supply for a critical project? It's in these pages. Maybe not exactly what you are looking for, but close enough to allow you to make modifications for it to be designed specifically for your application. How about an FSK modem? It will also be found in these pages along with AM/FM radios, range detection devices, super high-power amplifiers (and super low-powered ones as well), oscillators, generators, and many more. Some are highly practical for a myriad of applications. Others are highly specialized. Some don't seem to have any practical application at all! It depends on what you need and how you can adapt it.

While the main criterion for choosing these circuits was to present a well-rounded conglomeration, the practical aspect was also considered. Most of these circuits can be built inexpensively from components that are readily available. A brief description accompanies each installation and may provide an overview of circuit operations, tips on modifications or component substitution, suggested uses, the author's opinions, or all of these.

It is hoped that this book will prove to be a thorough and immediate reference source for the many circuits needed by technicians and engineers as they go about their pursuits of ever better and more efficient circuit designs.

Robert J. Traister

Chapter 1

Amplifier Circuits

This chapter features many different circuits that fall into the very general category of amplifiers. To be a little more specific, most of these circuits address the audio frequency spectrum and range from true power amplifiers to preamps to mixers. Each was chosen for its suitability to a wide range of applications and for its adaptability to still others. Such circuits are ideal when designing public address systems, phono/stereo output systems for automotive or home use, professional-quality recording equipment, and many other categories of audio output devices.

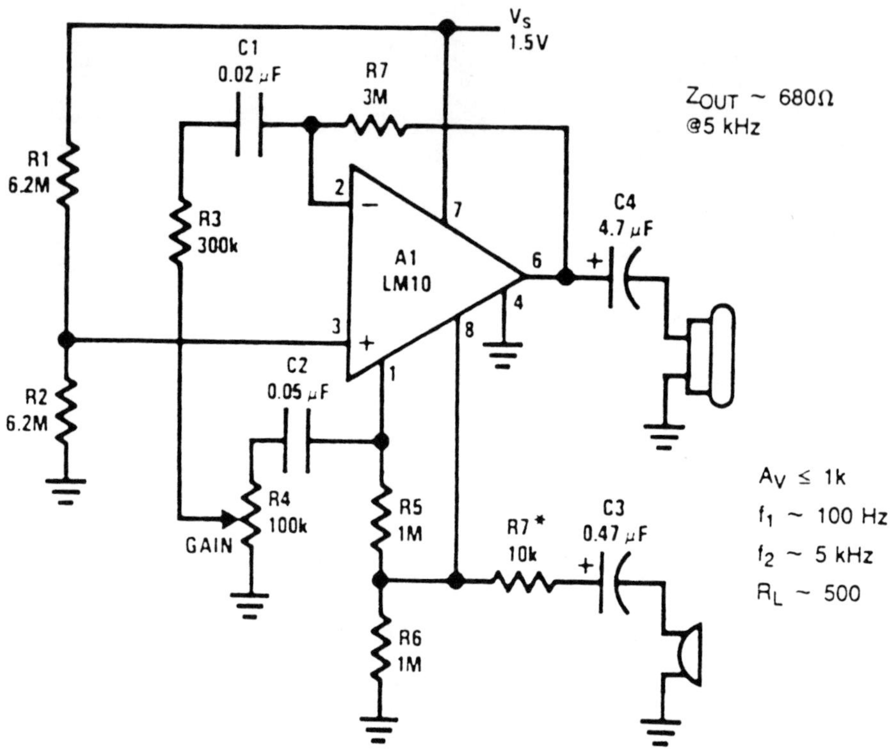

Microphone Amplifier. This op amp microphone amplifier is a self-contained public address system (with a very low power output). The reference, with a 500-kHz unity/gain bandwidth, is used as a preamplifier with a gain of 100. Its output is fed through a 100-kΩ gain/control potentiometer to the op amp section, which is connected for a gain of 10. Again, the LM10 has an op amp and a voltage reference on the same chip, and this is what allows for this unusual configuration. The combination gives a 60-dB gain with a bandwidth of 10 kHz when unloaded and 5 kHz when loaded at 500 Ω. Input impedance is 10 kΩ. This concept can be carried only so far, however, as using the reference as a preamplifier in this fashion can cause excess noise. We get away with it here because the reference voltage is low and therefore the noise contribution is also low. Also, the signal swing at the reference output is strictly limited. It cannot swing much below 150 mV, nor closer than 800 mV to the supply potential.—"Linear Applications Databook." Reprinted with permission of National Semiconductor.

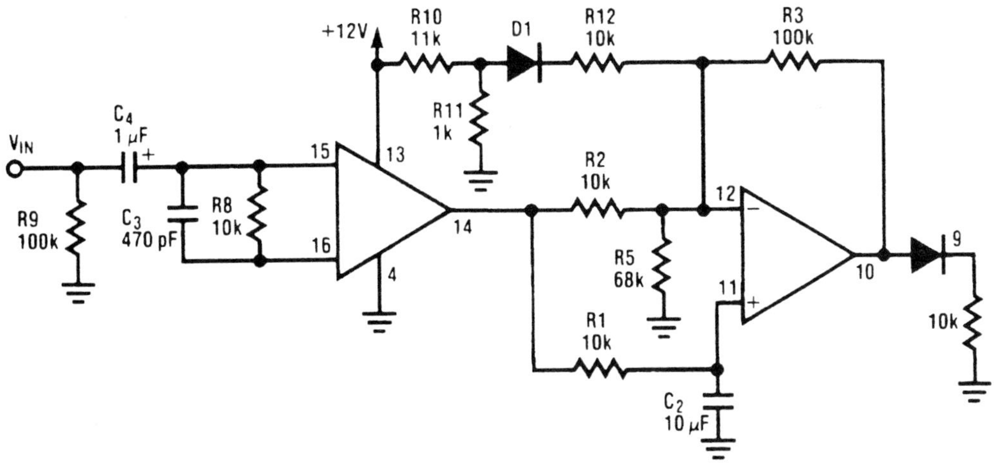

Microphone Preamplifier with Capacitor-Coupled Input. This circuit uses the National Semiconductor LM1897 LNP (low-noise preamplifier) IC, which is intended to be directly coupled to the signal source. Direct coupling permits faster turn-on and less low-frequency noise than is possible with capacitor-input systems, but there are many applications where it is mandatory that the signal source be referenced to ground and coupled through a capacitor. This circuit accomplishes this task and exhibits a flat frequency response suitable for microphone preamplifier purposes. R8 provides a DC path for input bias current, and its value should be as low as possible without loading the source. For this circuit 10 kΩ does the job nicely.—"Linear Databook III." Reprinted with permission of National Semiconductor.

Audio Mixer. With up to 4 input channels, this audio mixer uses a single NE542 dual low-noise preamplifier IC and 12 discrete components. This circuit is easily miniaturized and can be constructed in a very short time and with a minimum of expense. The single-polarity 12-V DC power supply is most convenient, as it allows for easy blending with other audio equipment operating from the same supply voltage. The 1-μF capacitors and the 500-kΩ variable resistors should be high-quality types in order to prevent noise from being induced into the system.—"Linear Data Manual Volume 1: Communications." Courtesy of Signetics Company, a division of North American Philips Corporation.

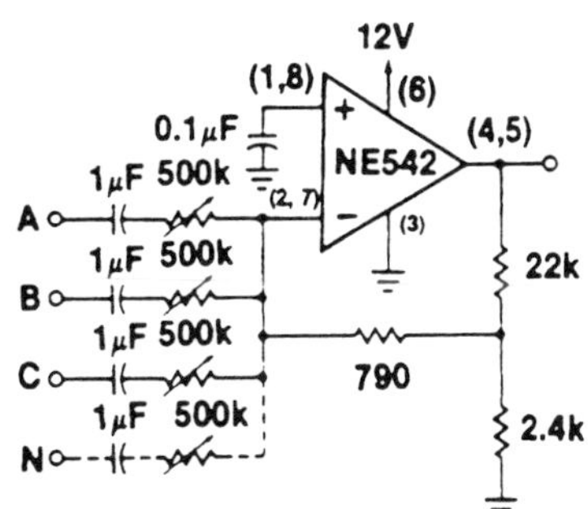

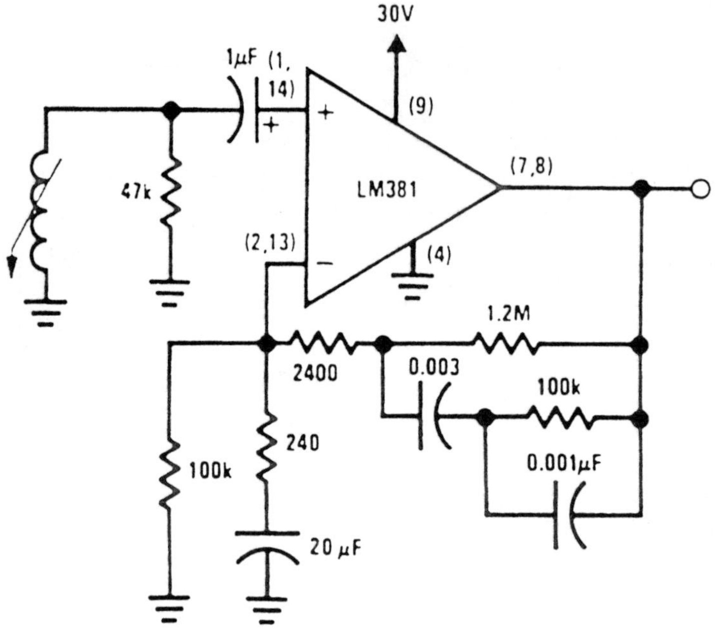

Magnetic Phono Preamplifier. Driven by the magnetic cartridge in a turntable, this circuit provides adequate output to drive moderate-power audio amplifiers. While designed for 30-V DC operation, it should operate at a 24-V DC potential with little degradation in performance. This preamplifier is designed to allow magnetic cartridge devices to utilize amplifier circuits that are normally designed to be driven directly by crystal or ceramic cartridges.—"Linear Databook III." Reprinted with permission of National Semiconductor.

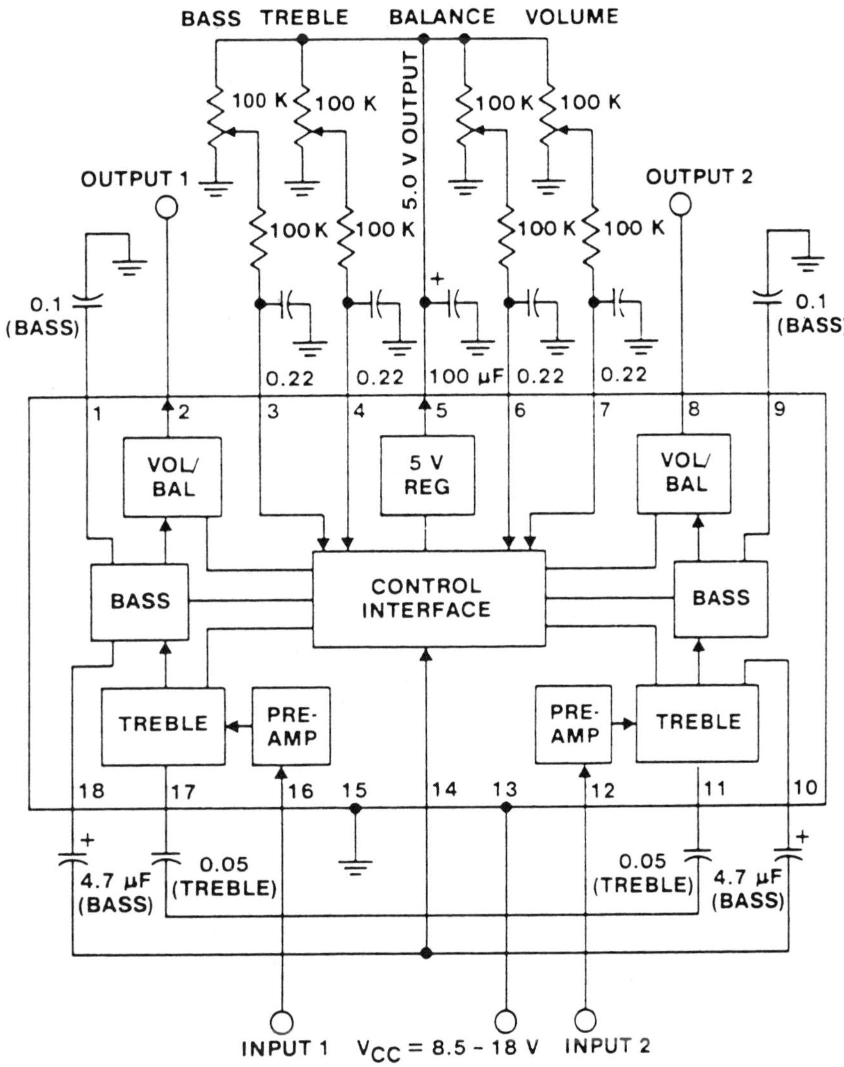

Stereo Sound Volume/Tone Control. The CA3259 stereo volume/tone control IC allows much of the complexity involved in designing such a circuit from discrete components to be avoided entirely. The circuit shown will operate from any supply providing from about 9 to 18 V DC. Simple DC inputs allow volume/tone control to be effected by four inexpensive potentiometers or even a remote control system. This chip offers distortion characteristics of approximately 0.1% and a better than 45-dB channel separation. Bass and treble responses for this circuit are defined by a single capacitor per control per channel. All in all, this is a compact, inexpensive, and very useful circuit that can be built for a minimum of cost and construction time. Outputs are low impedance.—"RCA Solid State Databook—Integrated Circuits for Linear Applications." Courtesy of Harris Semiconductor.

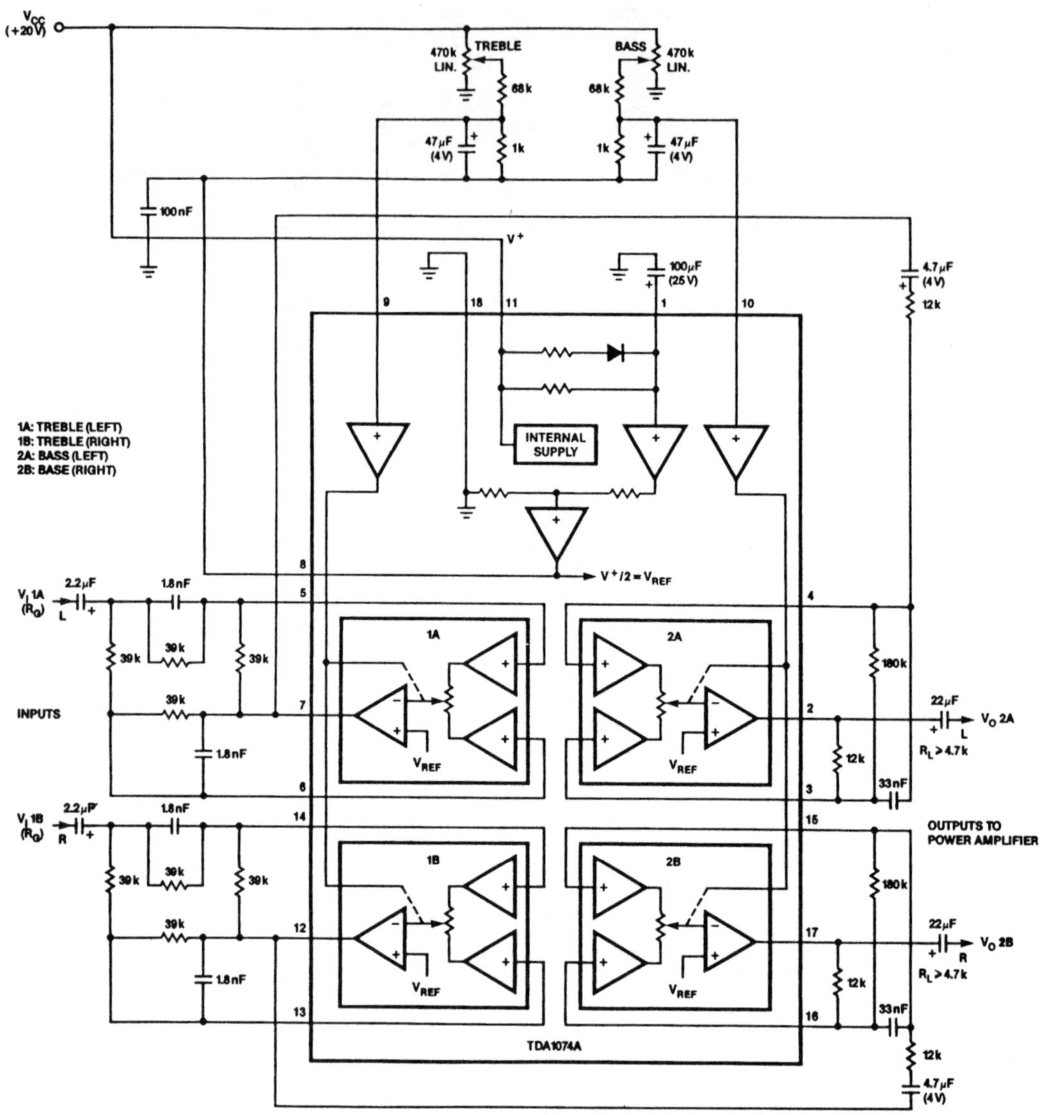

Treble/Bass Control for Stereo Applications. The TDA1074A is an IC designed specifically for use as a volume and tone control circuit in stereo amplifiers. The dual tandem potentiometer IC consists of two ganged pairs of electronic potentiometers with the eight inputs connected via impedance converters. The four outputs drive individual operational amplifiers. Z1 through Z4 are 22-kΩ carbon resistors. R_g is equal to 60 Ω while R1 is 4.7 Ω. The coupling capacitors at the inputs and outputs are C1 = 2.2 μF and C0 = 10 μF.—"Linear Data Manual Volume 1: Communications." Courtesy of Signetics Company, a division of North American Philips Corporation.

Tape Playback Amplifier. Accepting its input from the magnetic head of a tape player deck, this circuit produces a 500-mV output. The Signetics NE542 low-noise stereo amplifier IC is used here with a bare minimum of discrete components (four resistors and three capacitors). A single-polarity 12-V DC supply is all that is needed, although power supplies offering 9 V DC should work almost as well.—"Linear Data Manual Volume 1: Communications." Courtesy of Signetics Company, a division of North American Philips Corporation.

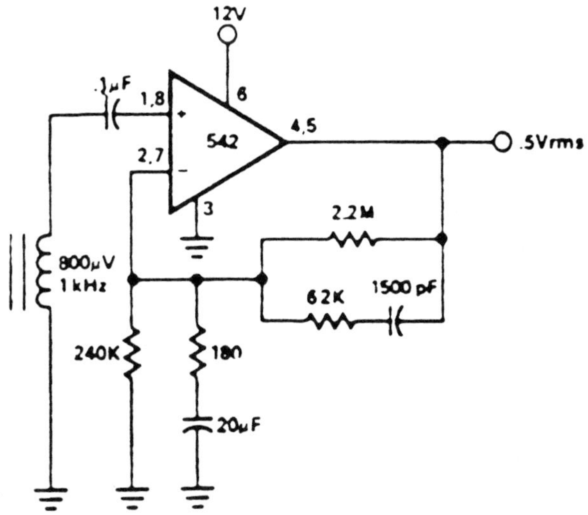

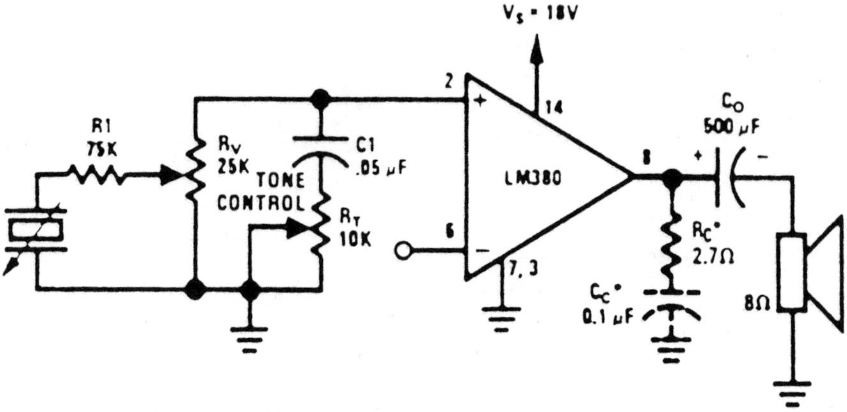

Phonograph Amplifier. This simple circuit offers a single-channel phono amplifier with tone and volume controls. The volume control is a voltage divider type while the tone control utilizes a high-frequency roll-off RC circuit connection. The RC circuit shown in dotted lines at the output of this amplifier is for RF suppression and attenuates oscillations in the 5- to 10-MHz region. This portion of the circuit need not be included if the amplifier will not be used around RF-sensitive circuitry.—"Linear Applications Databook." Reprinted with permission of National Semiconductor.

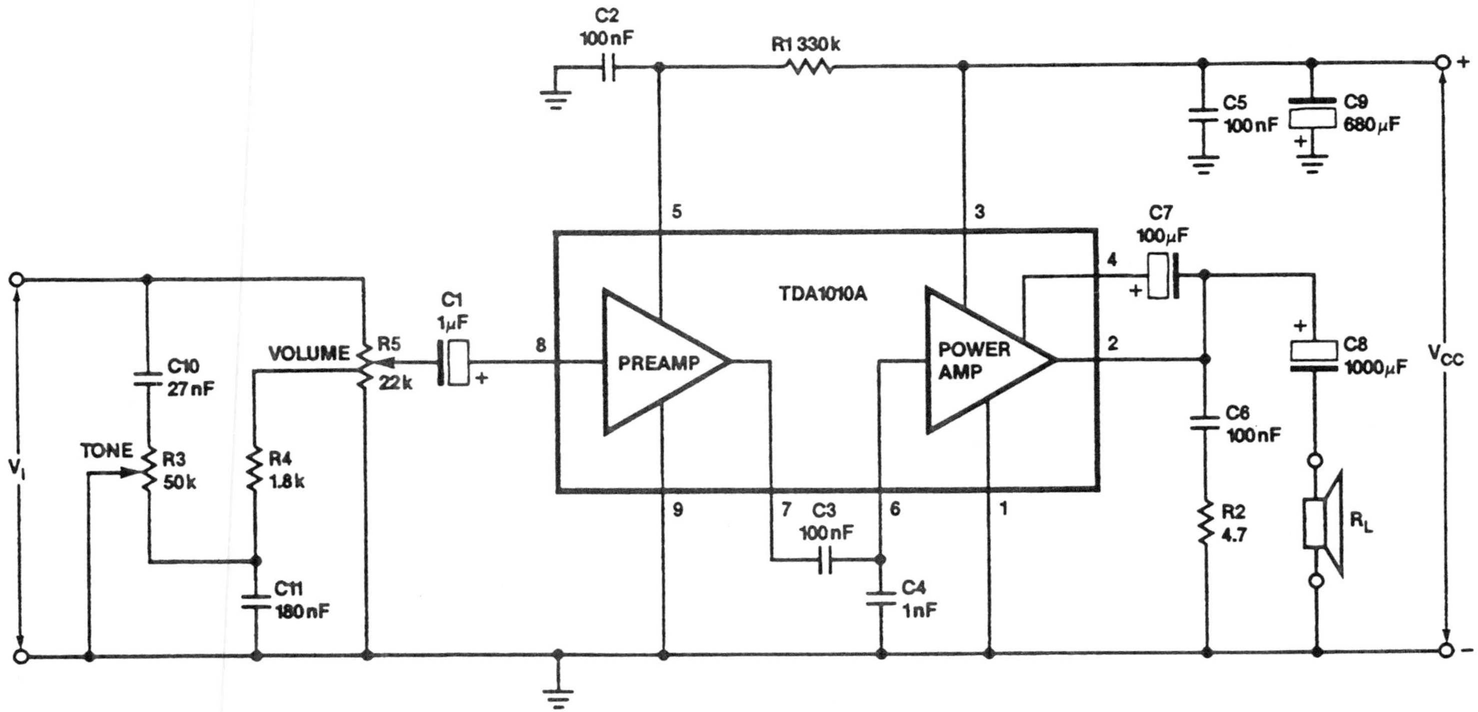

6-Watt Audio Amplifier for Radios. This circuit will boost the output of a small transistor radio to a full 6 W into a 4-Ω speaker. The heart of the circuit is the Signetics TDA1010A audio amp/preamp IC. The input is taken directly from the earphone jack of the radio and fed to the preamp portion of the IC via a tone/volume control network. Output from the power amplifier is capacitively coupled to the speaker. All components are standard and component value is mostly noncritical. Supply voltages can vary but must be kept at or below 22 V DC. The higher supply voltages will result in a true 6-W output. Power drops to about 3 W out at 12 V DC.—"Linear Data Manual Volume 1: Communications." Courtesy of Signetics Company, a division of North American Philips Corporation.

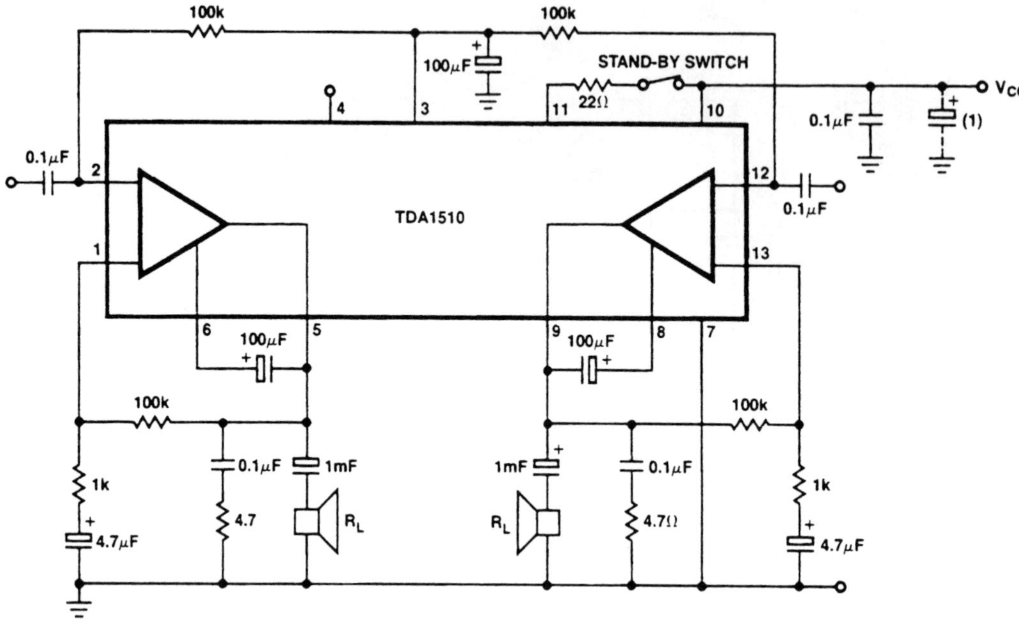

6 Watts per Channel Stereo Amplifier. The Signetics TDA1510 audio amplifier IC used in this circuit is a monolithic integrated class B output amplifier developed primarily for automobile stereo applications. It possesses two audio amplifiers on a single chip, each of which is capable of a maximum 12-W output (to a 2-Ω load). In this configuration, the output power will be about 6 W into 4-Ω speakers. Operating from the automobile electrical system, this circuit is straightforward, but component placement is critical and minimum lead lengths are absolutely essential in any type of automotive device to diminish the possibility of induced noise from the electrical system.—"Linear Data Manual Volume 1: Communications." Courtesy of Signetics Company, a division of North American Philips Corporation.

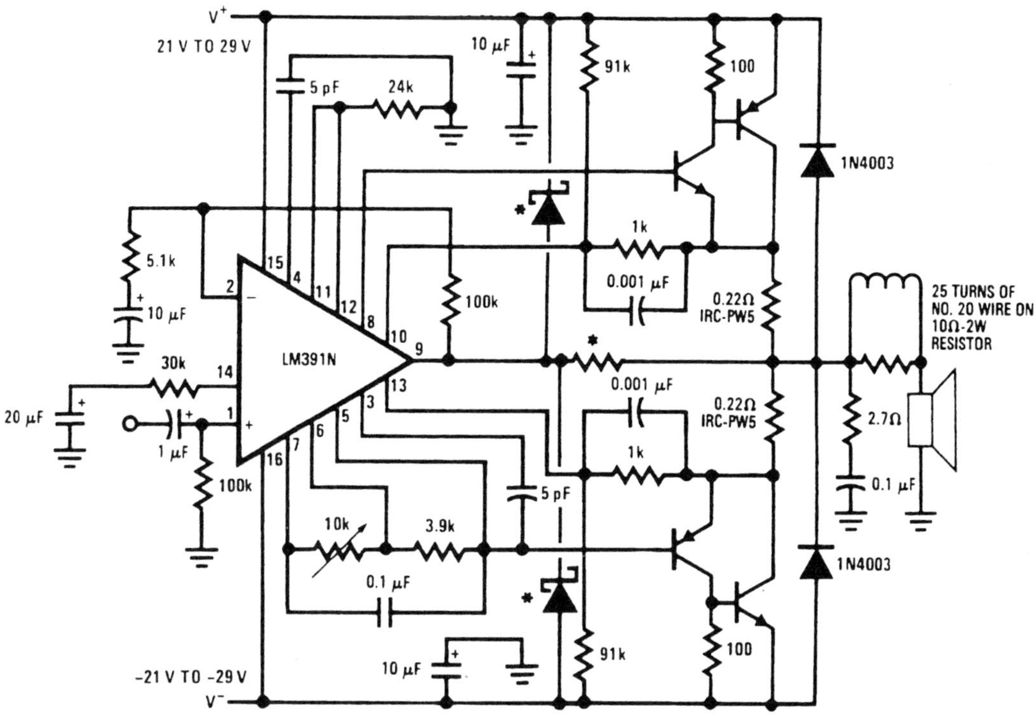

*Additional protection for LM391N; Schottky diodes and R ≅ 100Ω.

20-Watt Amplifier with 1-Second Turn-On Delay. With an output power of 20 W into an 8-Ω load, this circuit offers plenty of audio power in a simple design. It does require a bipolar power supply in the 24-V DC range, which is a bit of an inconvenience when compared with designs that offer unipolar operation. However, the overall operational characteristics of this circuit are worth the trouble. The transistor units are MJE711 and MJE721 complementary driver types. This circuit will output a full 30 W into a 4-Ω load.—"Linear Databook III." Reprinted with permission of National Semiconductor.

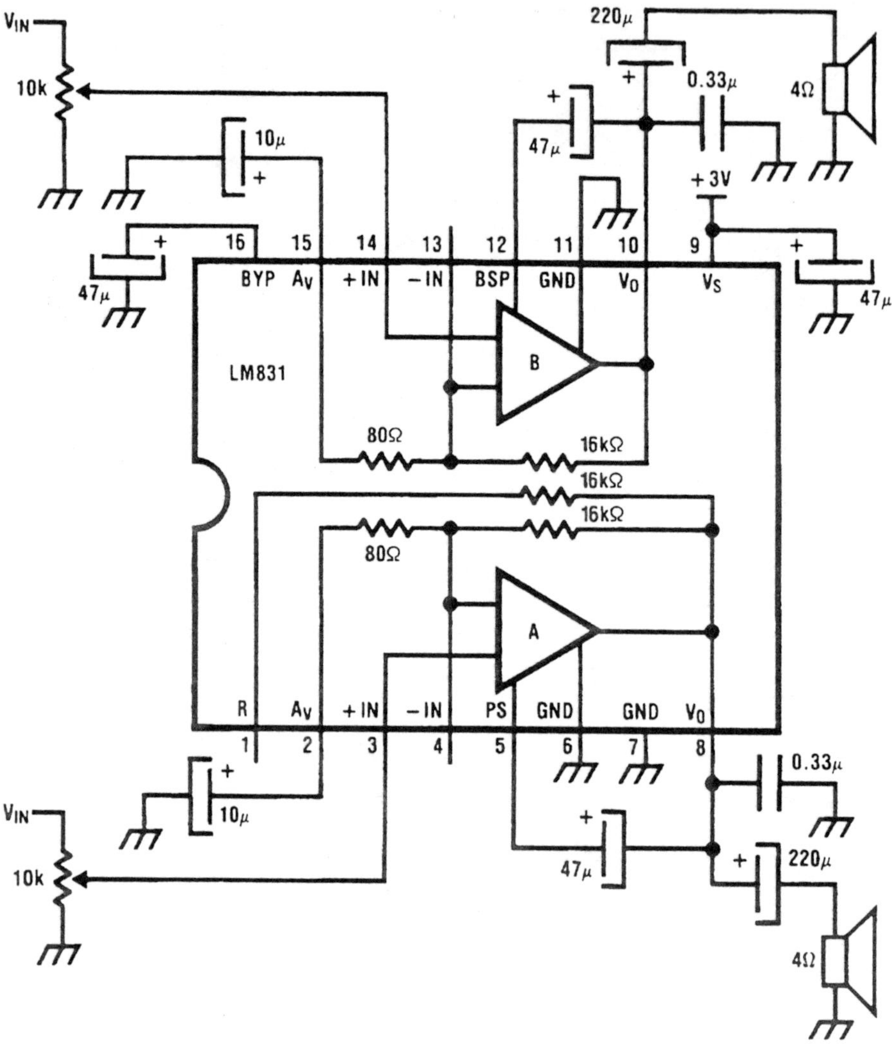

Low-Voltage Audio Power Amplifier. Operating from a power supply of only 3 V DC, this amplifier is capable of outputting approximately 220 mW to a 4-Ω load. The heart of this circuit is the National Semiconductor LM831 LVAPA (low-voltage audio power amplifier) IC which exhibits low-noise and low-distortion characteristics even at this voltage level. Note that the input is directly coupled to the IC, eliminating the usual coupling capacitor. This circuit has applications in the design and modification of portable tape recorders, portable radios, and other devices that are generally compact, hand carried, and must operate from small battery packs.—"Linear Databook III." Reprinted with permission of National Semiconductor.

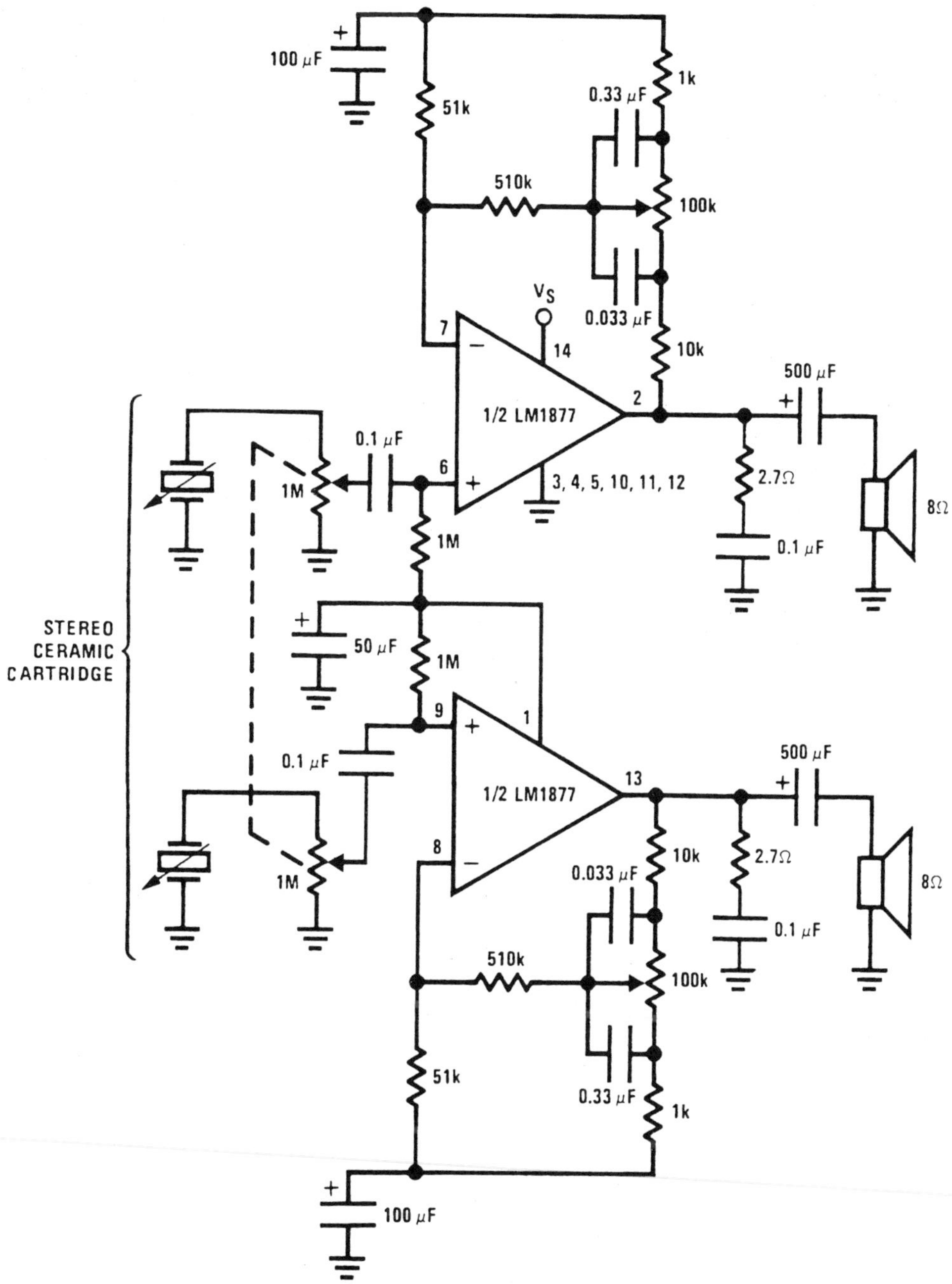

Stereo Turntable Amplifier. This two-channel stereo amplifier accepts input from a stereo ceramic cartridge and provides output to two 8 Ω speakers. A built-in bass control is provided by the two 100-kΩ variable resistors connected across the discrete outputs. This circuit is capable of driving each speaker with a full 2 W of audio power maximum when a 24-V DC power supply is used. This is a very versatile circuit in that it will accept supply voltages as low as 6 V DC. The ganged 1-MΩ variable resistors serve as a cross-channel volume control.—"Linear Databook III." Reprinted with permission of National Semiconductor.

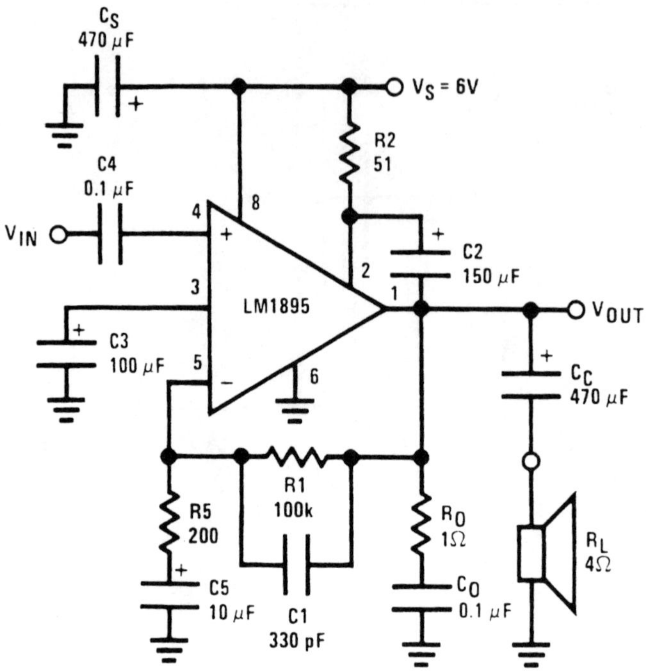

6-Volt Portable Amplifier. Intended for use in compact AM/FM radios, cassette players, and the general line of battery-operated devices, this amplifier has a bandwidth of approximately 5 kHz and an output power rating of nearly 250 mW into a 4-Ω speaker. It will draw about 500 mA at the 6-V DC design voltage. The NS LM1895 audio power amplifier IC incorporates a unique patented compensation scheme which results in lower noise, lower distortion, and less AM radiation than conventional designs.—"Linear Databook III." Reprinted with permission of National Semiconductor.

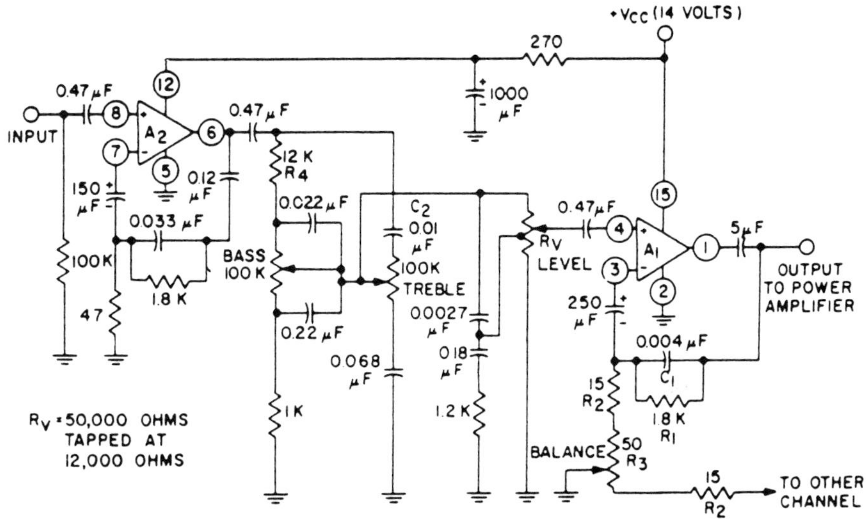

Performance Data

Gain at 1-kHz reference	47 dB
Boost at 100 Hz	11.5 dB
Boost at 10 kHz	11.5 dB
Cut at 100 Hz	10 dB
Cut at 10 kHz	9 dB

One-Half of a Stereo Amplifier. The unusual title for this circuit is simply one of convenience. It uses the CA3052 independent quad AC amplifier IC from RCA, which features four independent AC amplifiers. Two of these are used in each stereo preamplifier channel, so one IC is all that is required for a full two-channel preamplifier. However, this schematic shows only one-half of the total circuit, using two of the four available amplifiers. By doubling the external component complement shown here and wiring in the two remaining amplifiers, a full system can be had. The gain of each channel of this circuit, which operates from a 14-V DC supply, is approximately 47 dB at 1 kHz. It offers an 11.5-dB boost at 10 kHz to about 100 kHz. Total harmonic distortion is about 0.3%. As can be seen from the schematic, treble, bass, and balance controls are made a part of this configuration.—"RCA Solid State Databook—Integrated Circuits for Linear Applications." Courtesy of Harris Semiconductor.

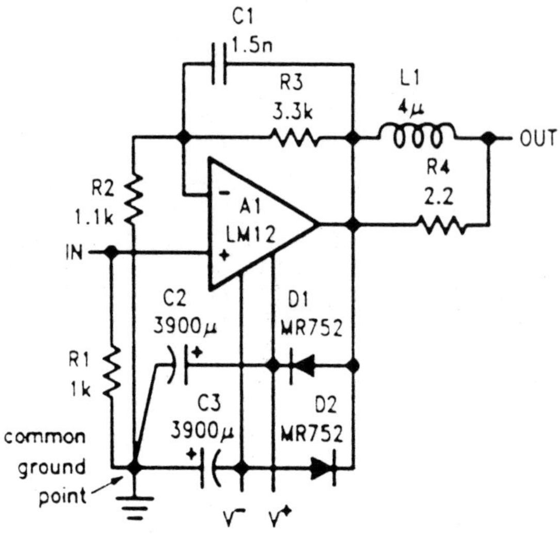

High-Power Audio Amplifier. This circuit uses the LM12 operational amplifier, which is a high-power device capable of delivering 150 W of sine wave power into a 4-Ω load with 0.01% distortion. Intermodulation distortion of this particular circuit has been measured at 0.015%. Using separate amplifiers to drive low- and high-frequency speakers gets rid of high-level crossover networks and attenuation. It also prevents clipping on the low-frequency channel from distorting the high frequencies. Naturally, this is a specialized op amp due to its high-powered nature. It is designed to operate at total supply potentials of 100 V.—"Linear Databook I." Reprinted with permission of National Semiconductor.

Son of Godzilla Op Amp Booster. The author doubts the practicality of this ▶ circuit from National Semiconductor, but the name makes it mandatory for inclusion in this text. Put simply, this is a very-high-voltage, high-current booster circuit that will allow an op amp to control up to 300 W for positive output potentials of up to 1000 V. This booster operates in a switching mode, so it exhibits very high efficiency, and it does not require a high-voltage power supply to achieve its high-voltage potential. The LM3524 switching regulator chip is used to pulse-width modulate the transistors. They, in turn, provide a 20-kHz drive to T1, a step-up transformer. The rectified and filtered output from T1 is fed back to the LF411 op amp which controls the input to the LM3524. In this manner, the high-voltage booster is controlled by the op amp feedback action. **Warning:** This circuit outputs a deadly potential many times higher than the lethal minimum. Extreme caution is mandated in circuit design, testing, and usage.—"Linear Applications Databook." Reprinted with permission of National Semiconductor.

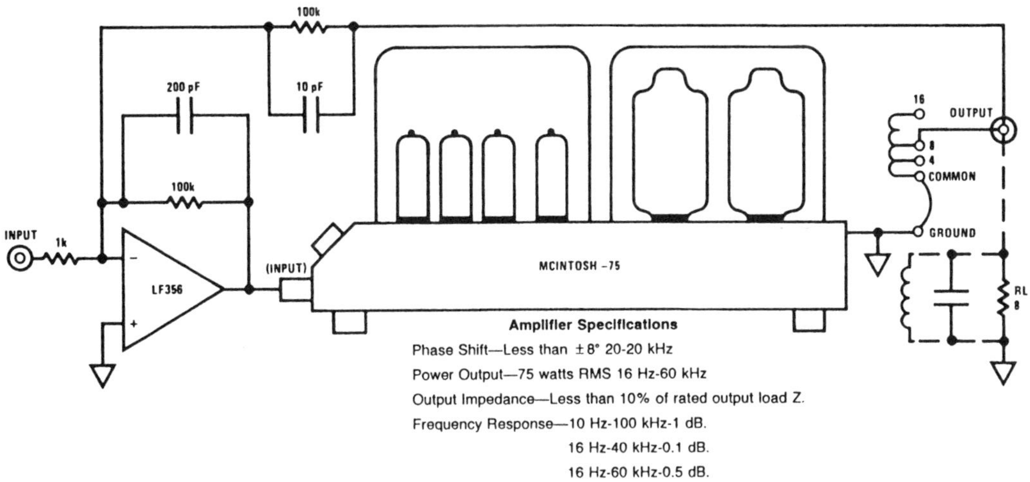
100k
200 pF
10 pF
100k
INPUT
1k
LF356
(INPUT)
MCINTOSH -75
16
OUTPUT
8
4
COMMON
GROUND
RL
8
Amplifier Specifications
Phase Shift—Less than ±8° 20-20 kHz
Power Output—75 watts RMS 16 Hz-60 kHz
Output Impedance—Less than 10% of rated output load Z.
Frequency Response—10 Hz-100 kHz-1 dB.
16 Hz-40 kHz-0.1 dB.
16 Hz-60 kHz-0.5 dB.

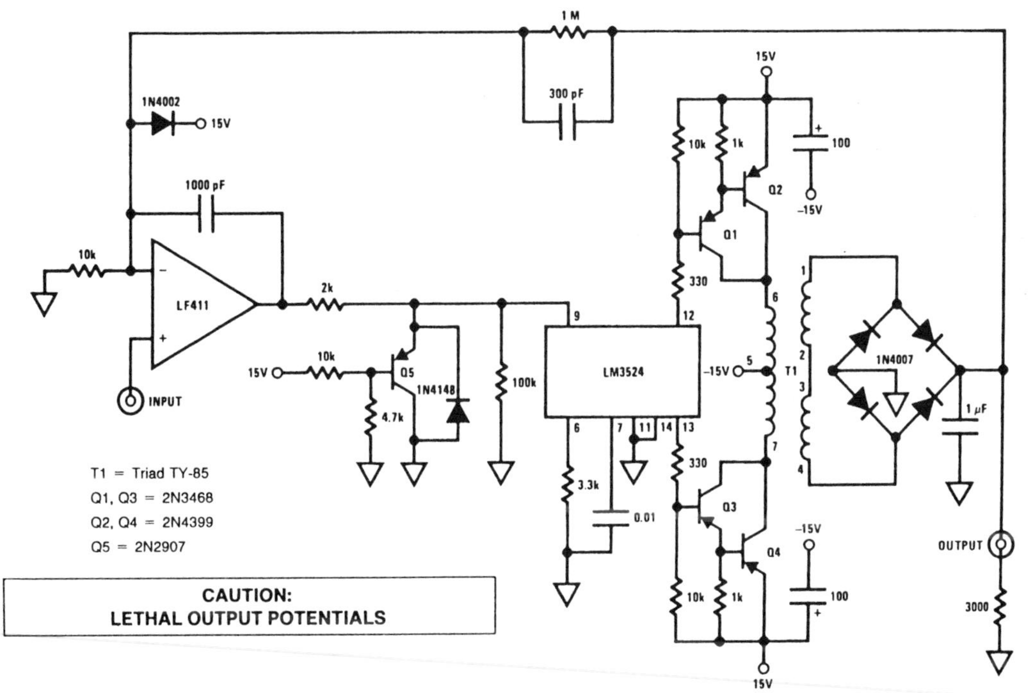
1 M
300 pF
1N4002
15V
1000 pF
10k
LF411
INPUT
2k
15V
10k
Q5
4.7k
1N4148
100k
LM3524
15V
10k
1k
100
Q2
-15V
Q1
330
-15V
T1
1N4007
1 μF
330
Q3
3.3k
0.01
Q4
-15V
10k
1k
100
15V
OUTPUT
3000
T1 = Triad TY-85
Q1, Q3 = 2N3468
Q2, Q4 = 2N4399
Q5 = 2N2907
CAUTION:
LETHAL OUTPUT POTENTIALS

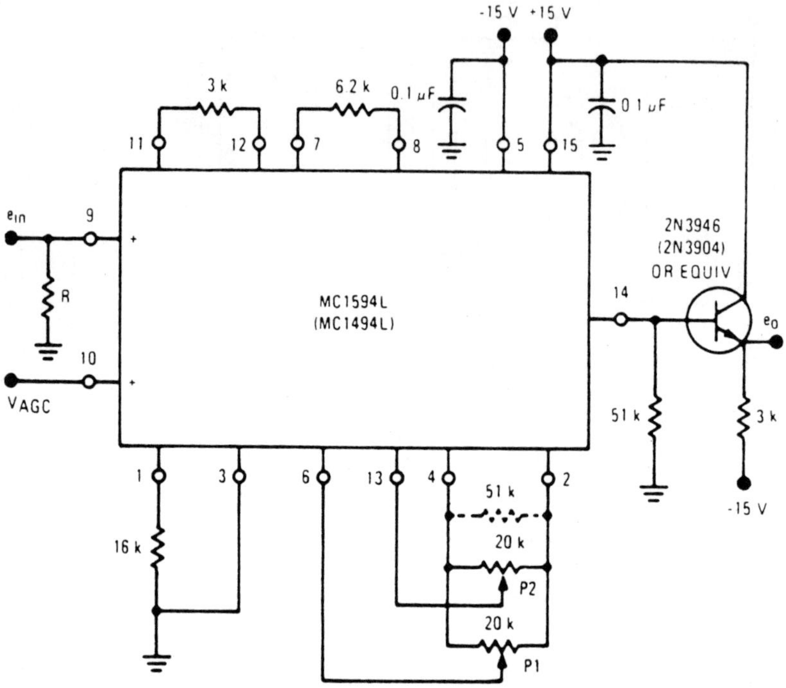

Wideband Amplifier with Linear AGC. This amplifier can handle 1 V (rms) and exhibits a gain of approximately 20 dB. The AGC operates via a 60-dB dynamic range with the application of an AGC voltage of from 0 to 1 V DC. The bandwidth of this amplifier is determined by the load resistor and the output stray capacitance. Owing to this, an emitter–follower buffer has been added to extend the bandwidth in excess of 1 MHz. The Motorola MC 1594L is a linear four-quadrant multiplier which allows this circuit to exhibit an AGC dynamic range which is theoretically infinite.—"Linear and Interface Integrated Circuits." Copyright of Motorola, Inc. Used by permission.

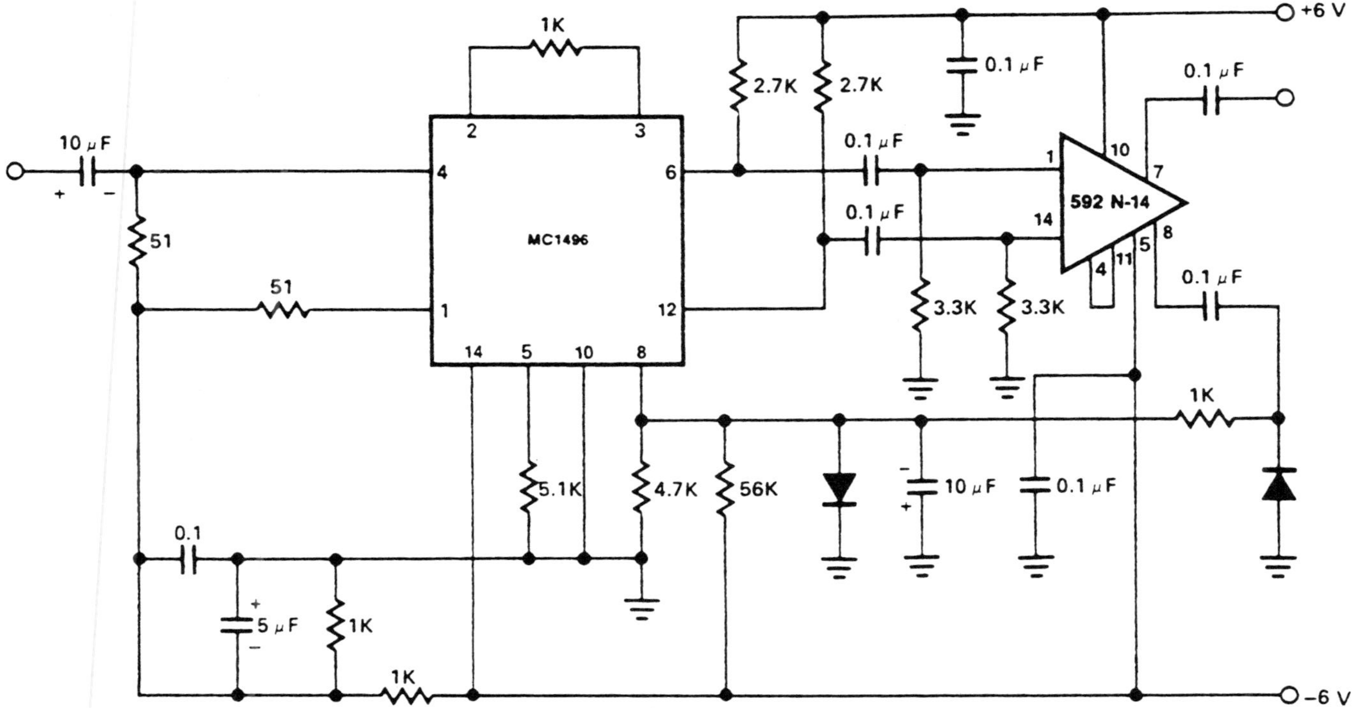

Wideband AGC Amplifier. This circuit connects the Signetics NE592 video amplifier IC in conjunction with the MC1496 balanced modulator to form an excellent automatic gain control system. The signal is fed to the signal input of the balanced modulator and RC-coupled to the NE592. Unbalancing the carrier input of the balanced modulator causes the signal to pass through unattenuated. Rectifying and filtering one of the video amplifier's outputs produces a DC signal which is proportional to the AC signal amplitude. After filtering, the control signal is applied to the balanced modulator, causing its gain to change. This circuit is designed to operate from a bipolar 6-V DC power supply and includes a component complement which is standard and relatively noncritical.—"Linear Data Manual Volume 1: Communications." Courtesy of Signetics Company, a division of North American Philips Corporation.

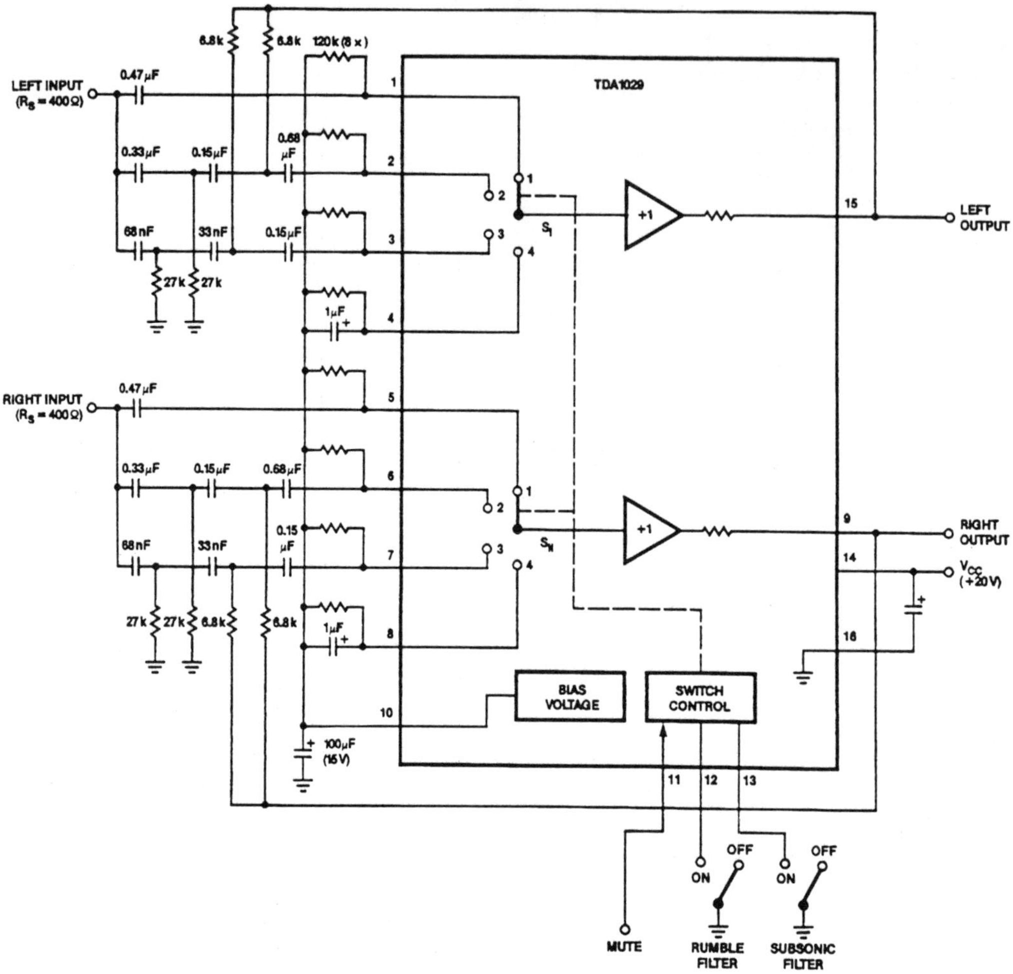

SWITCH CONTROL

FUNCTION	V_{11-16}	V_{12-16}	V_{13-16}
Linear	H	H	H
Subsonic filter 'on'	H	H	L
Rumble filter 'on'	H	L	X
Mute 'on'	L	X	X

Active High-Pass Filter for Stereo Applications. This circuit uses the Signetics TDA1029 stereo audio switch IC which consists of a dual operational amplifier connected as an impedance converter. Each amplifier has four mutually switchable inputs which are protected by clamping diodes. Input currents are independent of switch position and the outputs are short-circuit protected. The configuration shown here forms a third-order active high-pass filter with Butterworth response. It is a four-function circuit that can select mute, rumble filter, subsonic filter, and linear response. It operates from a single-polarity 20-V DC power supply.—"Linear Data Manual Volume 1: Communications." Courtesy of Signetics Company, a division of North American Philips Corporation.

500-Nanowatt ×10 Amplifier. The LM4250 programmable operational amplifier from National Semiconductor is a moderately unusual device, owing to its wide range of operating voltages and especially the lower end of this range. This device can operate from a bipolar power supply with a potential as low as 1.0 V DC. The configuration shown in this schematic is that of an amplifier that offers a gain of 10 and whose power consumption is only 500 nW. In addition to the IC, only four other components are required. Due to the nature of this IC, no frequency compensation is necessary. While the supply voltage indicated for this circuit is 1.0 V DC, it can also be driven by two flashlight batteries.—"Linear Databook I." Reprinted with permission of National Semiconductor.

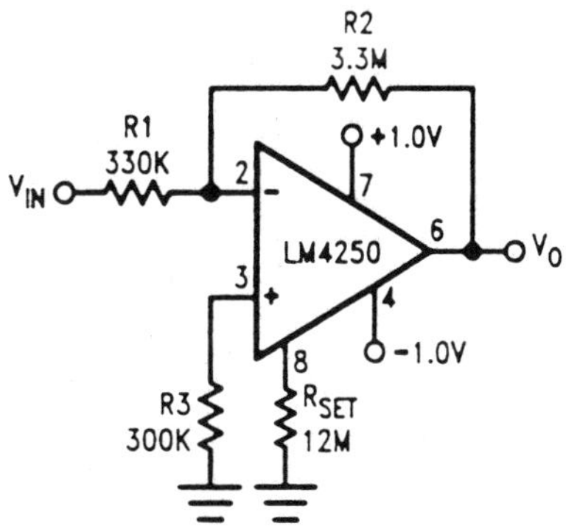

Variable Gain Circuit Using a Transimpedance Amplifier IC. A transimpedance amplifier is a current-to-voltage converter. The variable gain circuit consists of the aforementioned amplifier IC and a low-voltage op amp, the latter configured in a noninverting gain of five. The output drives the gate of the SD210 DMOS FET. The series resistance of the FET changes with this output voltage which, in turn, changes with the gain of the transimpedance amp. With the components shown, this circuit has a distortion figure of less than 1%.—"Linear Data Manual Volume 2: Industrial." Courtesy of Signetics Company, a division of North American Philips Corporation.

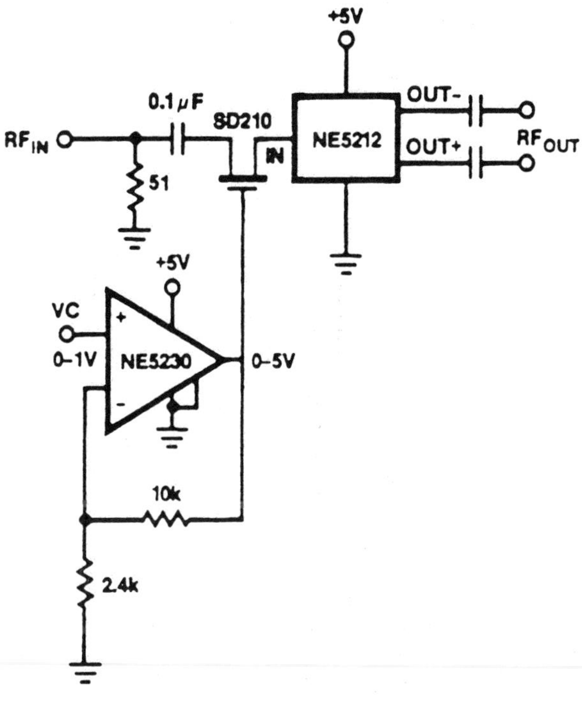

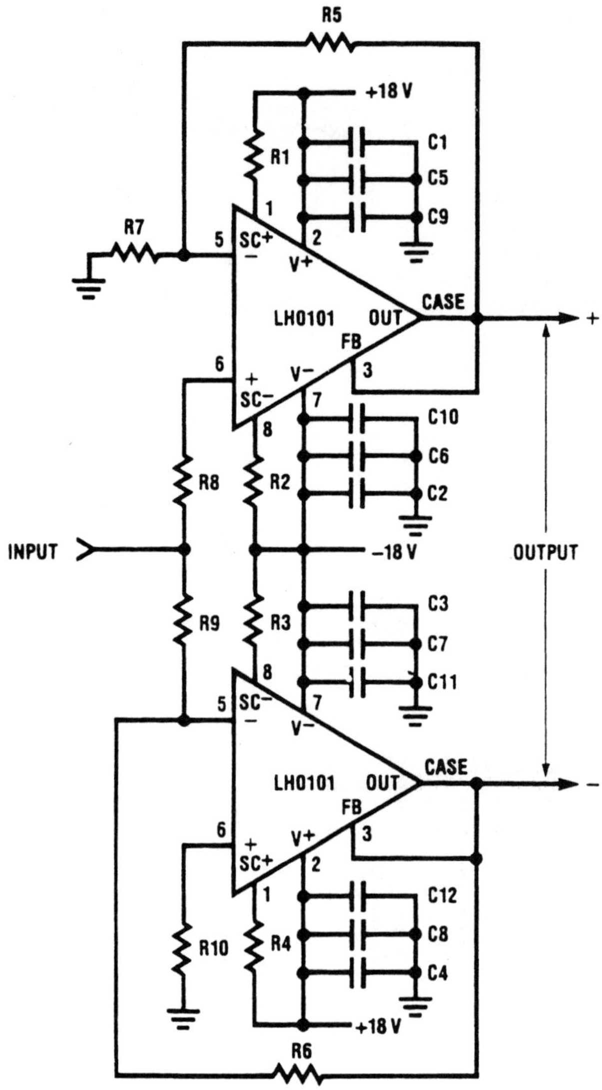

Bridge Audio Power Amplifier. This amplifier uses two LH0101 power op amps in a bridge configuration to arrive at a design that has an input impedance of 10 kΩ and a voltage gain of 3. It is capable of delivering 40 W of power into an 8-Ω load or a little more than half of that into 16 Ω. Total harmonic distortion at a full 40-W output is less than 0.15%. Intermodulation distortion at this same output is less 0.002%. The circuit operates from a single-polarity 18-V DC supply. Heat sink requirements are intense. It will be necessary to use a very large, finned heat sink or a smaller physical device with a fan or blower.—"Linear Applications Databook." Reprinted with permission of National Semiconductor.

R1–R4	Current Limit Resistor	0.15Ω, 2W
R5	Feedback Resistor	5 kΩ
R6	Feedback Resistor	15 kΩ
R7–R10	Input Resistors	10 kΩ
C1–C4	Bypass Capacitors	47 μF, 25V Electrolytic
C5–C8	Bypass Capacitors	10 μF, 25V Tantalum
C9–C12	Bypass Capacitors	0.1 μF, 25V Ceramic

Post-Amplifier for Op Amps. This circuit uses a CA3600 CMOS transistor array as a post-amplifier for a CA3080 op amp. Because the input impedance of the CMOS pair is comparatively high, the op amp operates under essentially unloaded conditions. Each CMOS pair can sink and source output current up to approximately 10 mA. The op amp output can also be directly coupled to bias the CMOS pair. This circuit produces a 130-dB gain, which is the sum of the approximate 30-dB gain in a single CMOS transistor pair and the 100-dB gain of the CA3080.—"RCA Solid State Databook—Integrated Circuits for Linear Applications." Courtesy of Harris Semiconductor.

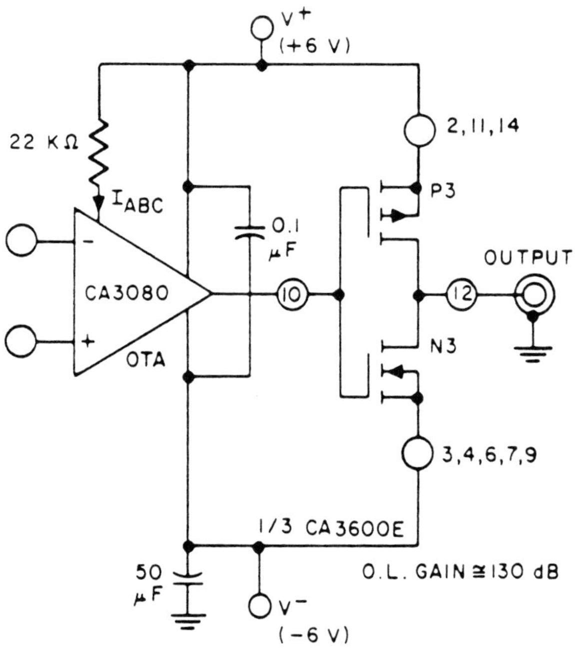

Chapter 2

Power Supply Circuits

The linear IC is mandatory in any modern type of regulated power supply circuit. Through linear IC design, highly regulated "rock solid" power supplies may be designed with a minimum of external components. Nor are such simplistic supply designs limited to only low-voltage/low-current applications. The versatility of the linear IC now makes the high-current or high-voltage regulated supply a comparatively inexpensive proposition while avoiding undue circuit complexity and building restraints.

The circuits featured here include the very simple, single IC, regulators, buck converters, DC-to-DC power supplies, various switching circuits, and several high-current designs. Each was chosen to represent an area in power supply design that has been made simpler and far more efficient due to continued development of the linear integrated circuit.

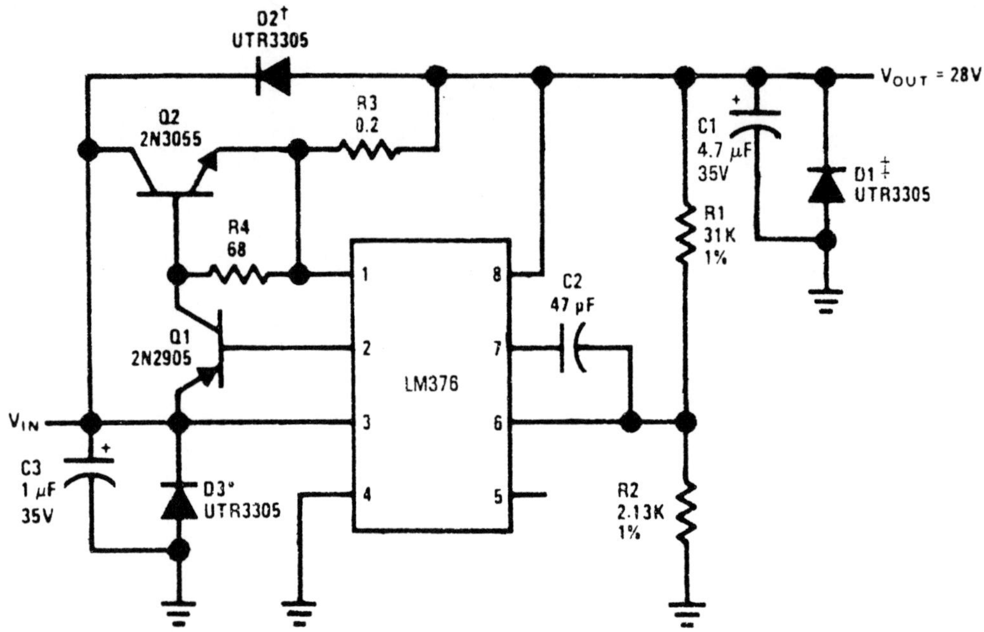

1.0-Ampere Regulator with Diode Protection. Diodes D1, D2, and D3 provide circuit protection in this 28-V DC, 1-A supply. D1 protects against output voltage reversal, while D2 provides protection from a shorted input or from inductive loads on the unregulated supply. D3 protects against input voltage reversal.—"Linear Databook I." Reprinted with permission of National Semiconductor.

5-Volt DC, 50-Ampere Line-Operated Supply. The Motorola MC3520 ▶ switchmode regulator control IC is used in this 20-kHz switching power supply. Within the input section, the 120-V AC line passes to a full-wave voltage doubler circuit comprised of CR1, CR2, C1, and C2. This provides 310 V DC to the power supply section of this circuit. R1 provides inrush surge current limiting. This resistor is shorted out of the circuit by Q1 after C1 and C2 are almost fully charged. Two switching transistors are used in the supply section of this circuit. Q2 and Q3 are wired in a half-bridge configuration to drive the high-frequency power transformer, T2. The output of T2 is rectified by Schottky diodes (CR9 and CR10). L1 and C4 make up the output filter network while L2 and C5 form a high-frequency filter to get rid of commutation spikes. The MC3520 provides pulse-width modulation (PWM) control for this supply. R2 is adjusted to obtain a frequency of 20 kHz. R3 is used to adjust the dead time of about 5 μsec per each half cycle. C6 provides a soft-start feature while Q4 ensures the soft-start after each system inhibit.—"Linear/Switchmode Voltage Regulator Handbook." Copyright of Motorola, Inc. Used by permission.

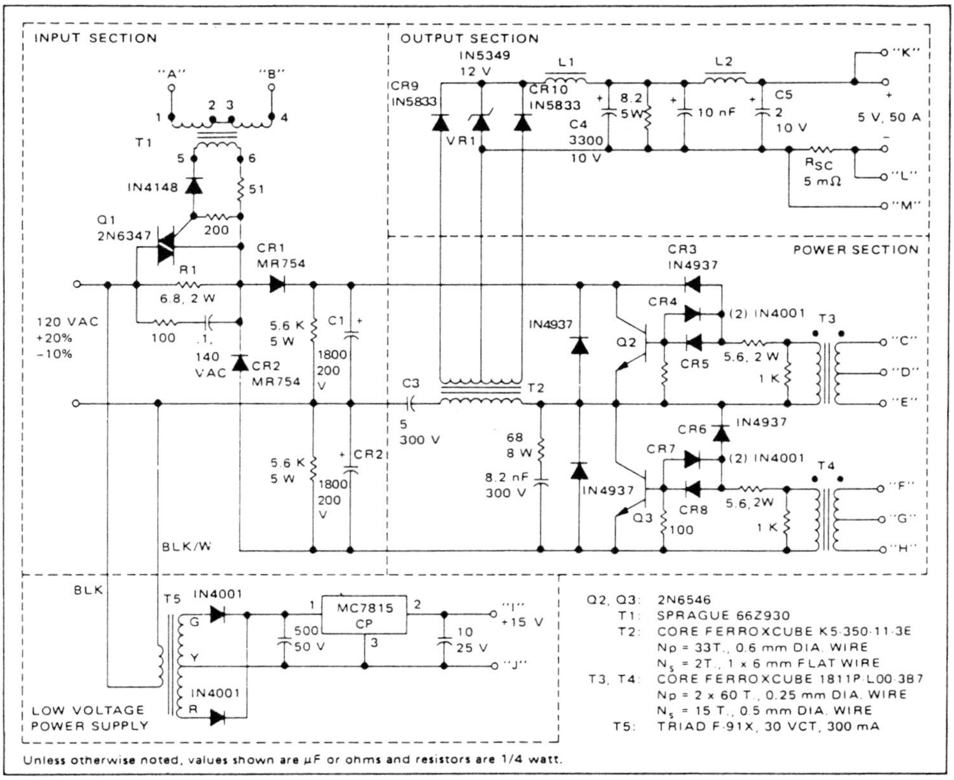
INPUT SECTION
OUTPUT SECTION
POWER SECTION
LOW VOLTAGE POWER SUPPLY
120 VAC +20% −10%
5 V, 50 A
Q2, Q3: 2N6546
T1: SPRAGUE 66Z930
T2: CORE FERROXCUBE K5-350-11-3E
Np = 33T., 0.6 mm DIA. WIRE
Ns = 2T., 1 x 6 mm FLAT WIRE
T3, T4: CORE FERROXCUBE 1811P-L00-3B7
Np = 2 x 60 T., 0.25 mm DIA. WIRE
Ns = 15 T., 0.5 mm DIA. WIRE
T5: TRIAD F-91X, 30 VCT, 300 mA
Unless otherwise noted, values shown are μF or ohms and resistors are 1/4 watt.

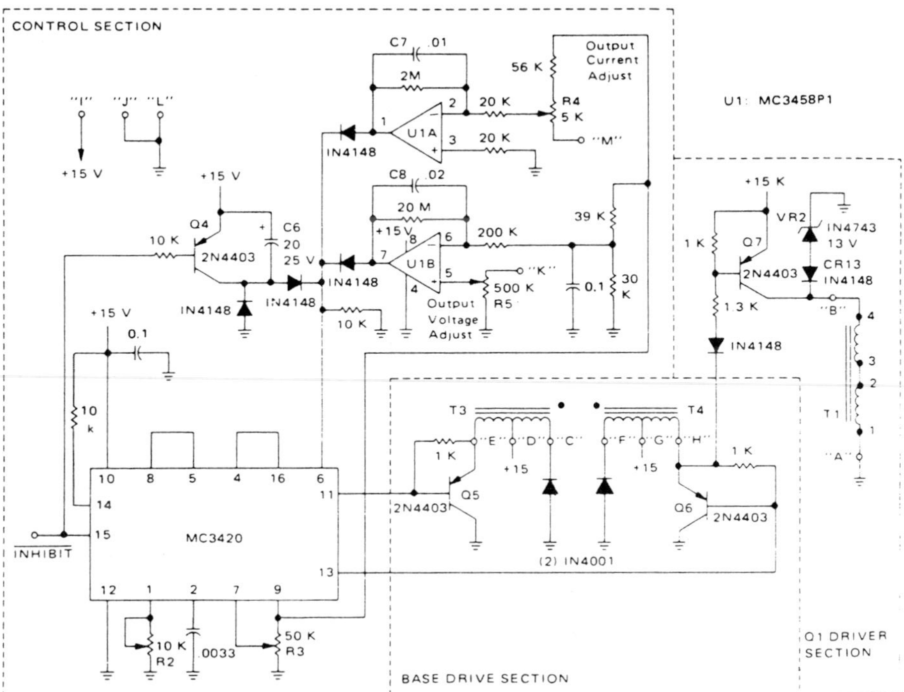
CONTROL SECTION
U1: MC3458P1
MC3420
Output Current Adjust
Output Voltage Adjust
INHIBIT
BASE DRIVE SECTION
Q1 DRIVER SECTION

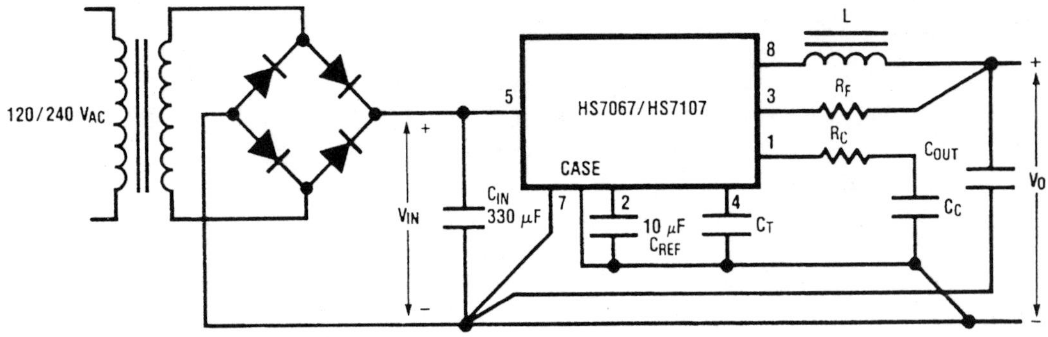

5-Volt DC Buck Converter. The buck converter is a very common configuration in switching power supplies. With this circuit a voltage may be stepped down with a bare minimum of discrete components. This supply accepts an input of 10 to 35 V DC and has a 5-V DC highly regulated output (load regulation is 40 mV). Output current for this circuit may be as high as 6 A, which is well within the 8-A maximum limits of the HS7067 series of switching regulators.—"Linear Databook I." Reprinted with permission of National Semiconductor.

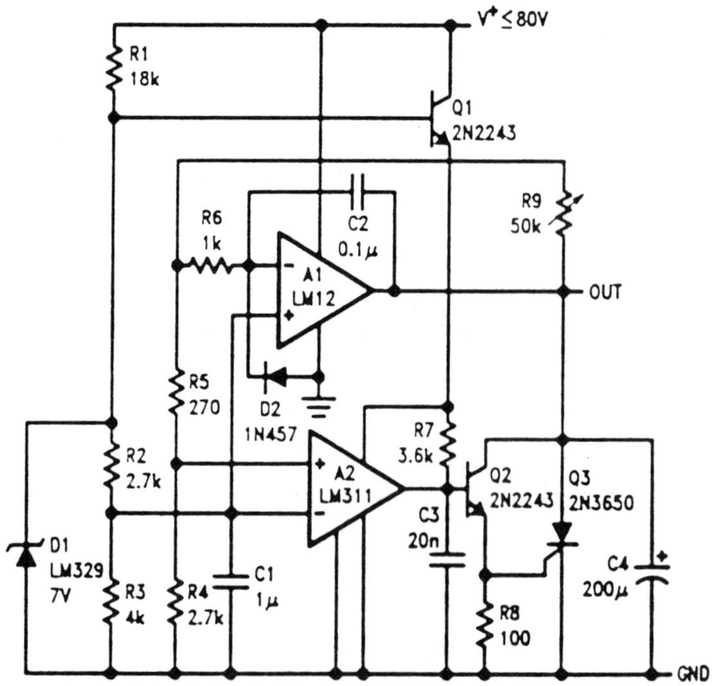

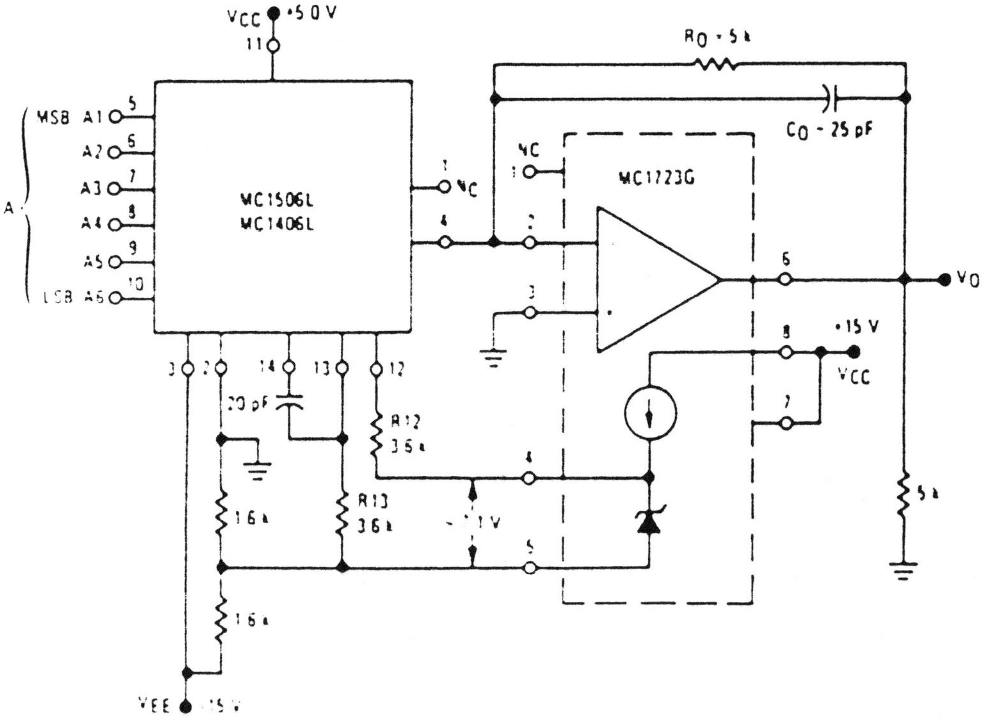

Programmable Power Supply. This digitally programmed power supply is capable of output voltage values of 0 to +6.3 V DC in 0.1-V DC increments (±0.05 V DC) or of 0 to 31.5 V DC in 0.5-V DC increments (±0.25 V DC).—"Linear and Interface Integrated Circuits." Copyright of Motorola, Inc. Used by permission.

◀ **Op Amp Voltage Regulator.** The op amp can be used as a positive or negative voltage regulator, and, when configured as such, the output can both source or sink current to absorb energy dumped back into the supply and prevent overvoltage with certain fault conditions. This characteristic is not exhibited by the more conventional dedicated voltage regulators. This circuit uses the LM12 high-voltage/high-power op amp. Its high-voltage capability allows regulated output voltages on the order of 70 V DC and also ensures start-up under worst-case, full-load conditions. This circuit has one input at ground and a reference current drawn from its summing junction. In this configuration, output voltage is proportional to the setting resistor, R2.—"Linear Applications Databook." Reprinted with permission of National Semiconductor.

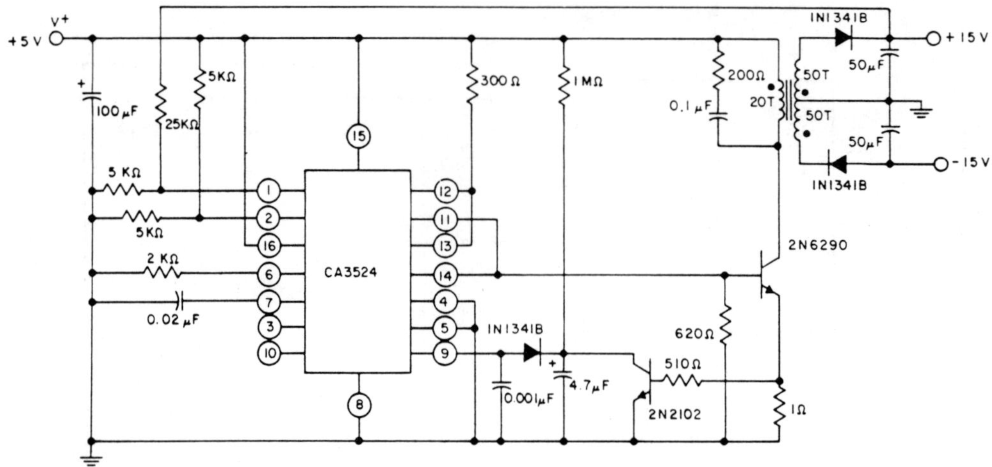

Dual Polarity 15-Volt Flyback Converter. Using a 5-V regulated line input, this circuit is able to produce a bipolar 15-V DC output at 20 mA. Reference voltage is provided by the input, and the internal reference generator in the CA3524 pulse-width modulator is unused. Current limiting in this circuit is accomplished by sensing current in the primary line and resetting the soft-start circuit. The transformer consists of a 20-turn primary and 100-turn center-tapped secondary wound on a Ferroxcube 2213P-A250-387 core or its equivalent.—"RCA Solid State Databook—Integrated Circuits for Linear Applications." Courtesy of Harris Semiconductor.

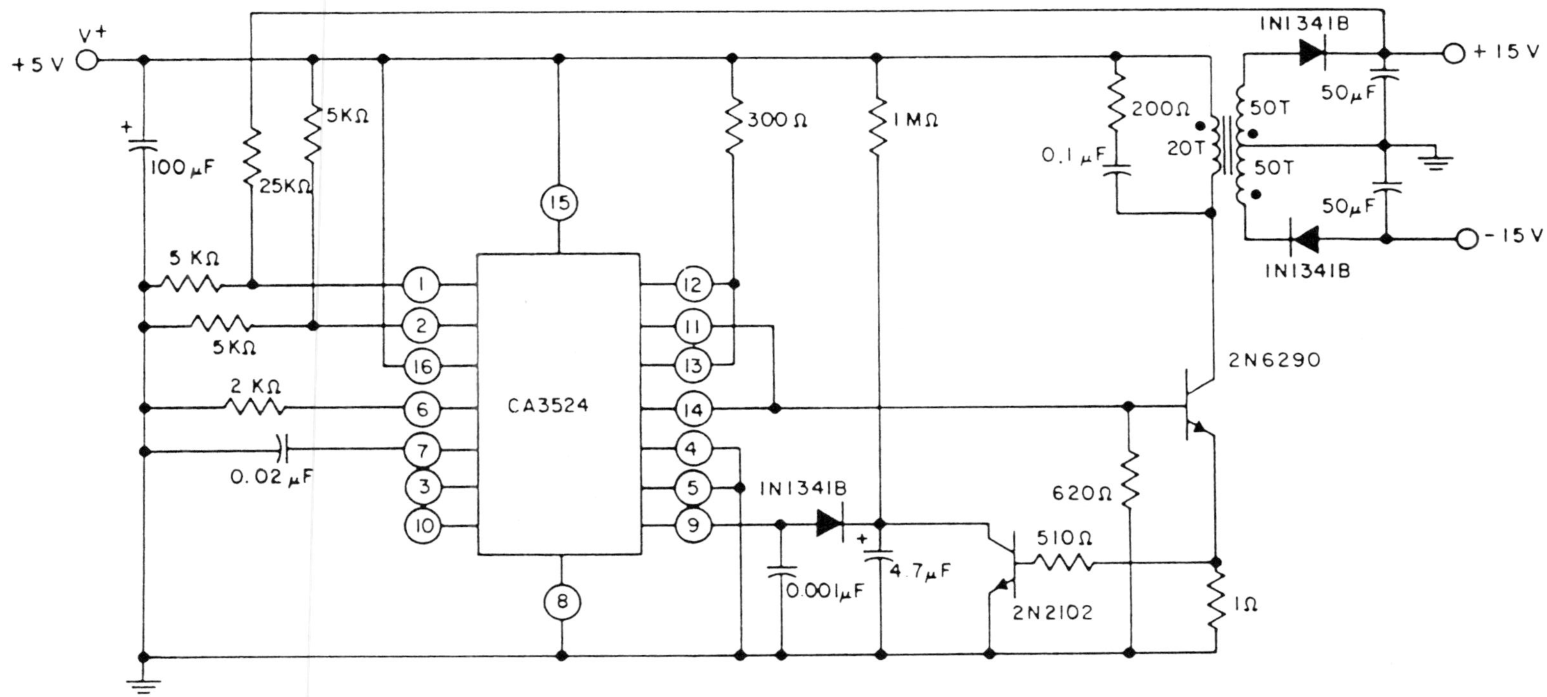

Variable Output Switching Power Supply. In this circuit, the RCA CA1524 regulating pulse-width modulator is used as a variable output power supply producing an output of 7 to 30 V DC at up to 3 A. By connecting the two output transistors in parallel, the duty cycle is doubled. As the reference voltage level is varied, the feedback voltage will track that level and cause the output voltage to change in accordance with the change in reference voltage.—"RCA Solid State Databook—Integrated Circuits for Linear Applications." Courtesy of Harris Semiconductor.

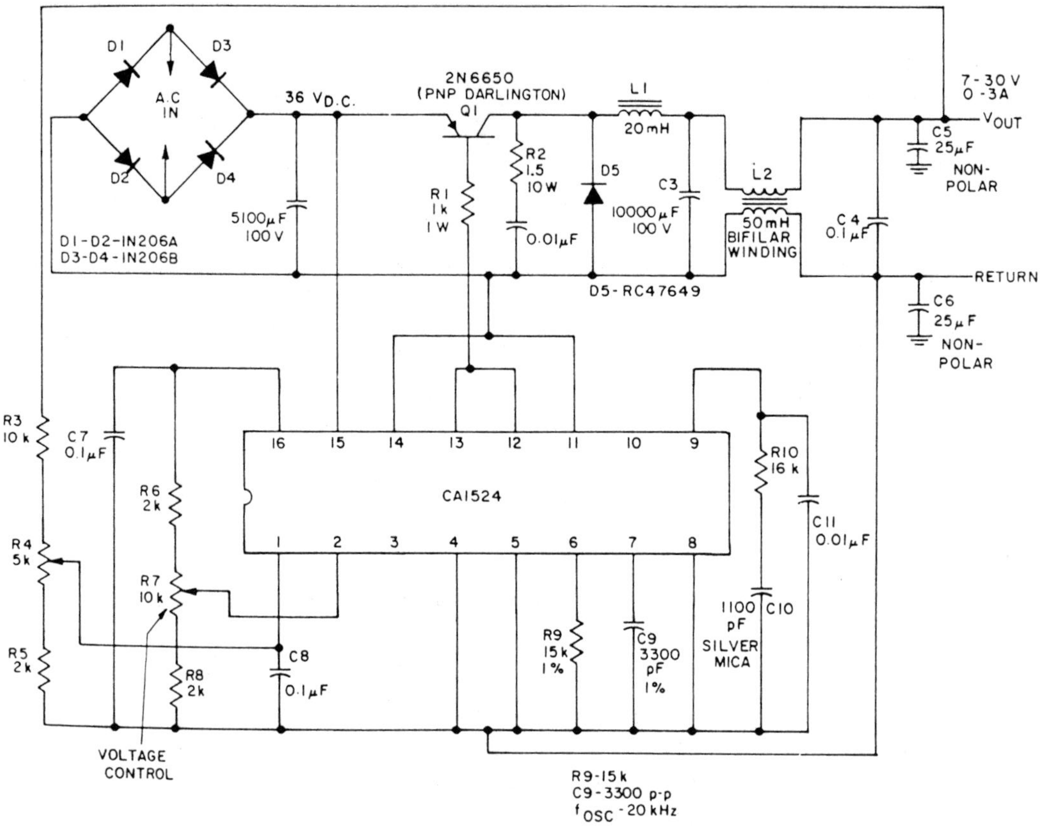

Flyback Converter. This circuit generates a dual 15-V DC output at 20 mA from a 5-V DC regulated line. Reference voltage is provided by the input and the internal reference generator is not used. The output stage of the CA3524 provides the drive for the two N-P-N transistors in this push/pull application. Since the internal flip-flop divides the oscillator frequency by two, the oscillator frequency must be set at twice that of the output frequency. Current limiting is accomplished in the primary winding of T1, so the pulse width will be reduced if transformer saturation occurs.—"RCA Solid State Databook—Integrated Circuits for Linear Applications." Courtesy of Harris Semiconductor.

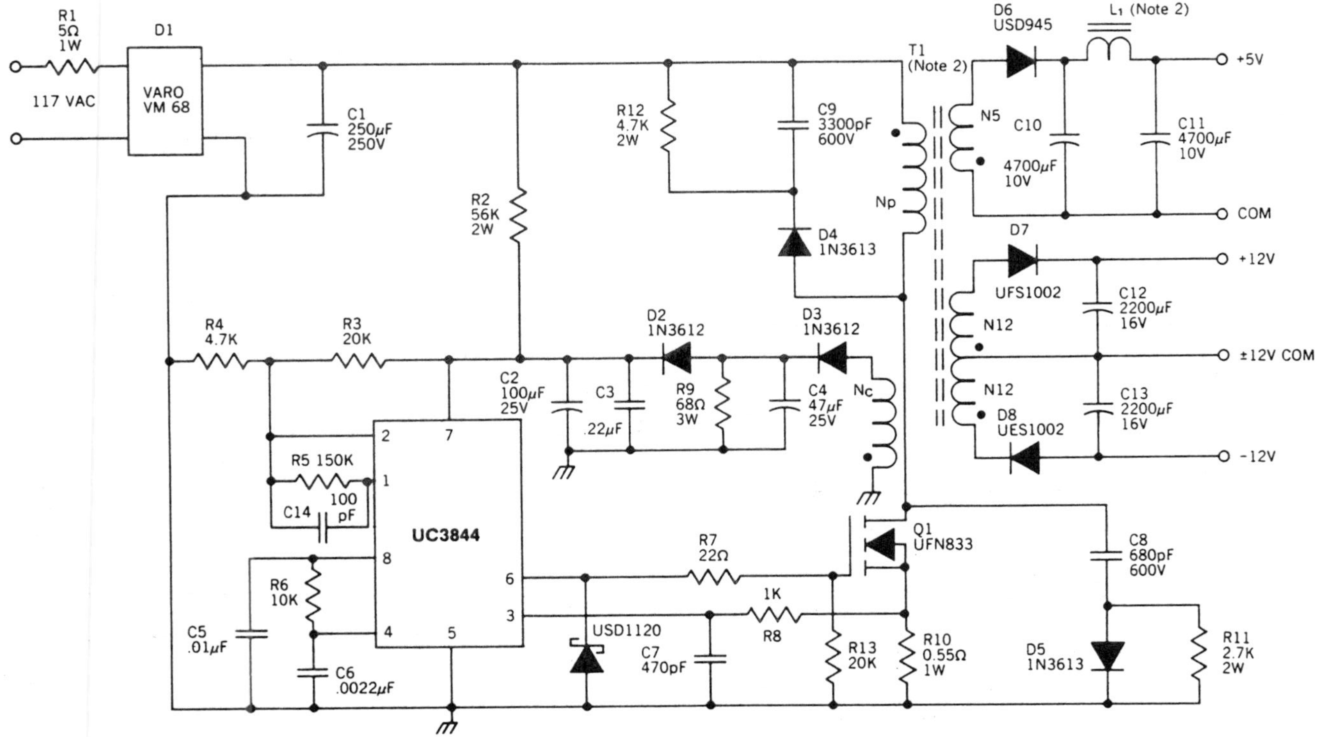

Inexpensive 25-Watt Off-Line Flyback Switching Regulator. This switching power supply is designed for applications requiring multiple output voltages. It can be constructed for less than $20, and the discontinuous mode flyback regulator used in this application provides good voltage tracking between outputs. This feature allows the use of primary side-voltage sensing, which is the key to the low-cost nature of the project, as the need for an isolated secondary feedback loop is eliminated. The input voltage can range from 95 to 130 V AC. Output ratings are +5 V DC (±5%) at 1 to 4 A; +12 V DC (±3%) at 100 to 300 mA; and −12 V DC at the same current rating. Switching frequency is 40 kHz. A small filter inductor (L1) is inserted in the 5-V secondary to reduce output ripple voltage to less than 50 mV. This inductor is also used to attenuate any high-frequency noise. It is comprised of four turns of AWG 18 wire on a Ferroxcube 204 T 250-3C8 toroidal core. The transformer is specially constructed from another Ferroxcube core. Its design is explained in the second set of schematics.—"Linear Integrated Circuits Databook." Courtesy of Unitrode Integrated Circuits.

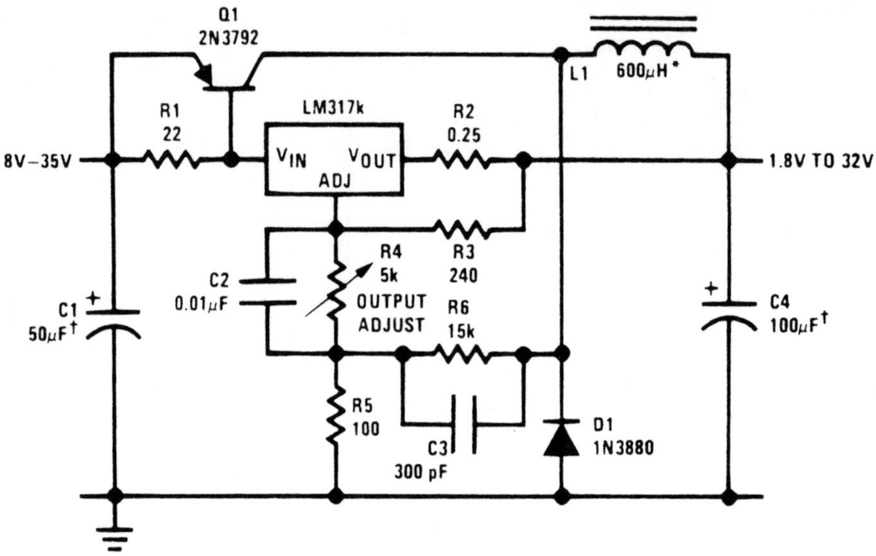

Adjustable 3-Ampere Switching Regulator. With an 8- to 34-V DC input, this circuit will provide an output of from 1.8 to 32 V DC, which is adjustable using the 5-kΩ potentiometer. Wind the inductor on an Arnold Engineering A-254168-2 core using approximately 60 turns of small gauge (#22) enamel clad wire.—"Linear Databook I." Reprinted with permission of National Semiconductor.

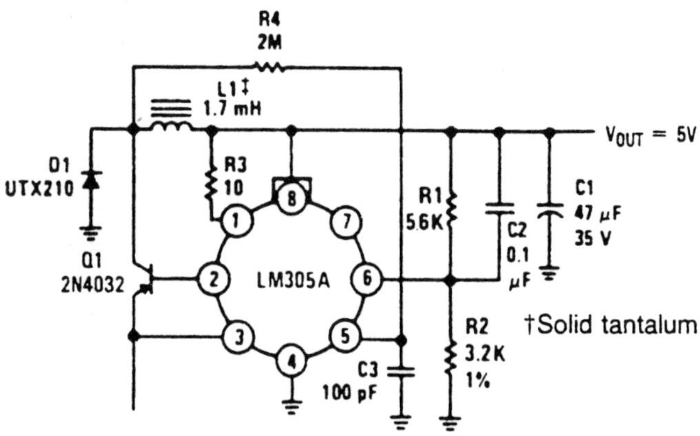

Switching Regulator. Providing a 5-V DC output with an input of 8.5 V DC, this circuit offers excellent regulation characteristics and uses more or less standard components. L1 consists of 125 turns of #22 enamel wire on an Arnold Engineering A-262123-2 core.—"Linear Databook I." Reprinted with permission of National Semiconductor.

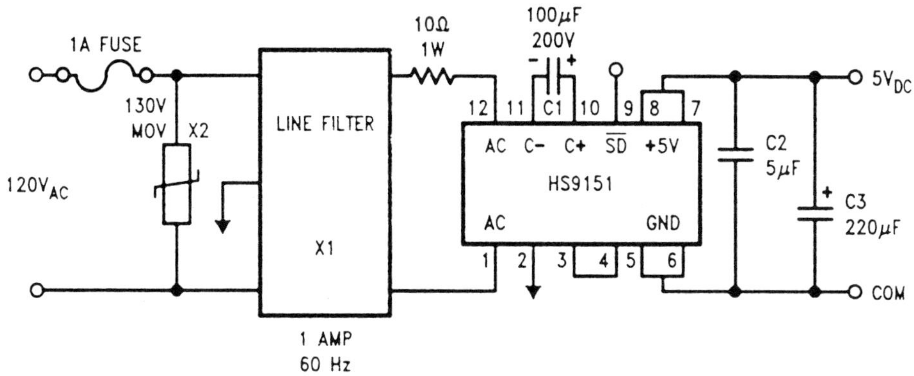

X1 SAE F15209 Line Filter, TRI-MAG G3-1 or equivalent
X2 GE V130LA1 Metal Oxide Varistor
C1 Mepco/Electra 3476 LK101M200 JMBS or equivalent
200V 100 μF Aluminum Electrolytic
C2 Rel-Cap PPMF505KIR OR Electronic Concepts 5MC 22B 505K
100V, 5 μF Polypropylene 50V, 5 μF Polycarbonate
C3 Mepco/Electra 3481 CE221V025 JDBS or equivalent
25V, 220 μF Aluminum Electrolytic

5-Volt DC, 3-Ampere Switching Power Supply. This circuit uses a hybrid power converter linear IC to form a very efficient and very simplistic power supply that delivers 3 A of current at 5 V DC. Here, a 100-μF input capacitor (between pins 10 and 11) provides a hold-up time of 16 msec. The 220-μF capacitor in parallel with the 5-μF polypropylene capacitor guarantees loop stability under highly varying line and load conditions.—"Linear Databook I." Reprinted with permission of National Semiconductor.

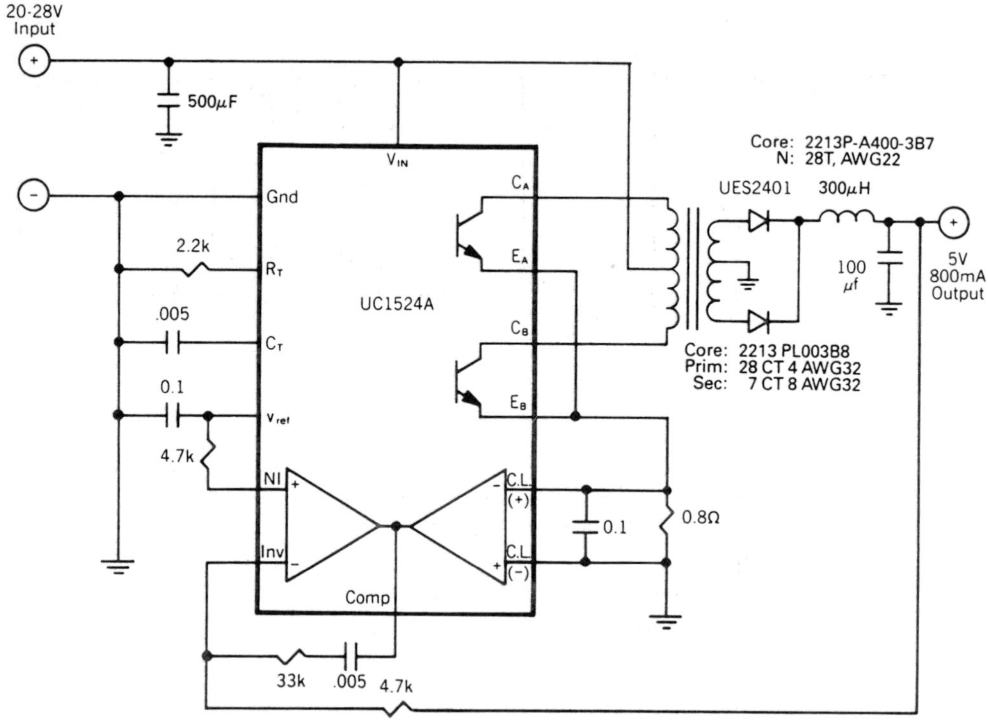

50-Watt Forward Converter Power Supply. Offering the advantages of the forward converter power supply configuration, this circuit overcomes some of the problems also associated with this design. The problem of isolated feedback control is solved by sampling the 5-V DC output level at the switching frequency by means of the 2N2222 transistor and transformer T2. During every switching cycle, the output voltage is transferred from N1 to N2 where it is peak detected to generate a primary-referenced signal to drive the PWM error amplifier. Operating at 40 kHz, this circuit provides an isolated 50 W of power with an efficiency of 83%, a high degree of regulation, and fast overload protection.—"Linear Integrated Circuits Databook." Courtesy of Unitrode Integrated Circuits.

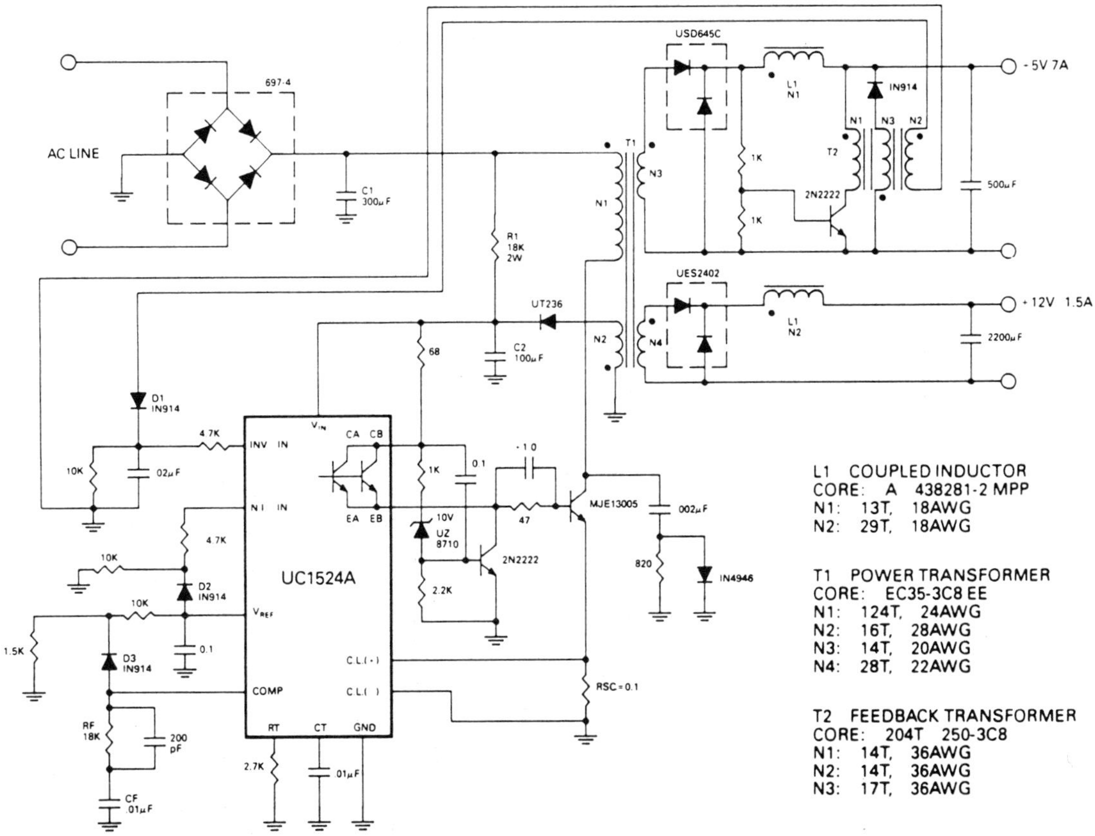

4-Watt DC-to-DC Converter without External Switching Transistors. Using the UC1524A PWM control IC, it is quite practical to construct a 4-W, highly regulated power supply without resorting to discrete switching transistors. Due to the higher voltage rating of the output transistors on this chip, the need for additional output transistors is not there, unless higher-power applications are intended. This circuit is a push/pull configuration that imposes a voltage of twice the supply across the "off" transistor. With a 28 : 7 turns ratio in the transformer, a 5-V, 800-mA output is achieved with 78% efficiency. Notice that only a few external parts are required, an unusual occurrence for any power supply of this type.—"Linear Integrated Circuits Databook." Courtesy of Unitrode Integrated Circuits.

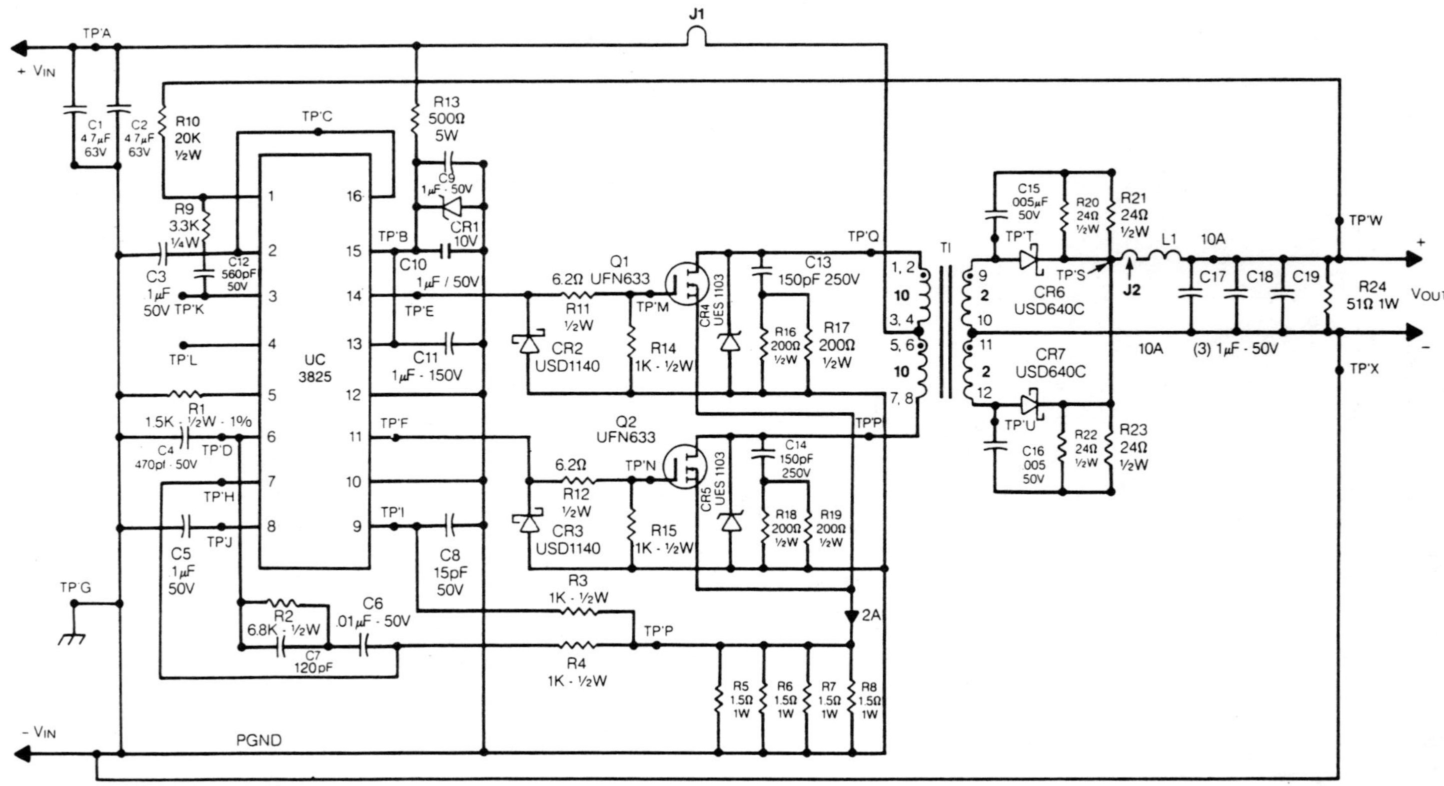

1.5-Megahertz Current Mode IC-Controlled 50-Watt Power Supply. A push/pull configuration is used for this DC-to-DC converter which is designed for a 42- to 56-V DC input and an output of 5 V DC at 10 A. The beneficial increase in switching speed (1.5 MHz) and dynamic performance is made possible by the Unitrode UC3825 pulse-width modulator IC. The transformer is a highly specialized unit that is not commercially available but must be hand wound using combined bundles of conductors (consult "Unitrode Linear Integrated Circuits Databook")—"Linear Integrated Circuits Databook." Courtesy of Unitrode Integrated Circuits.

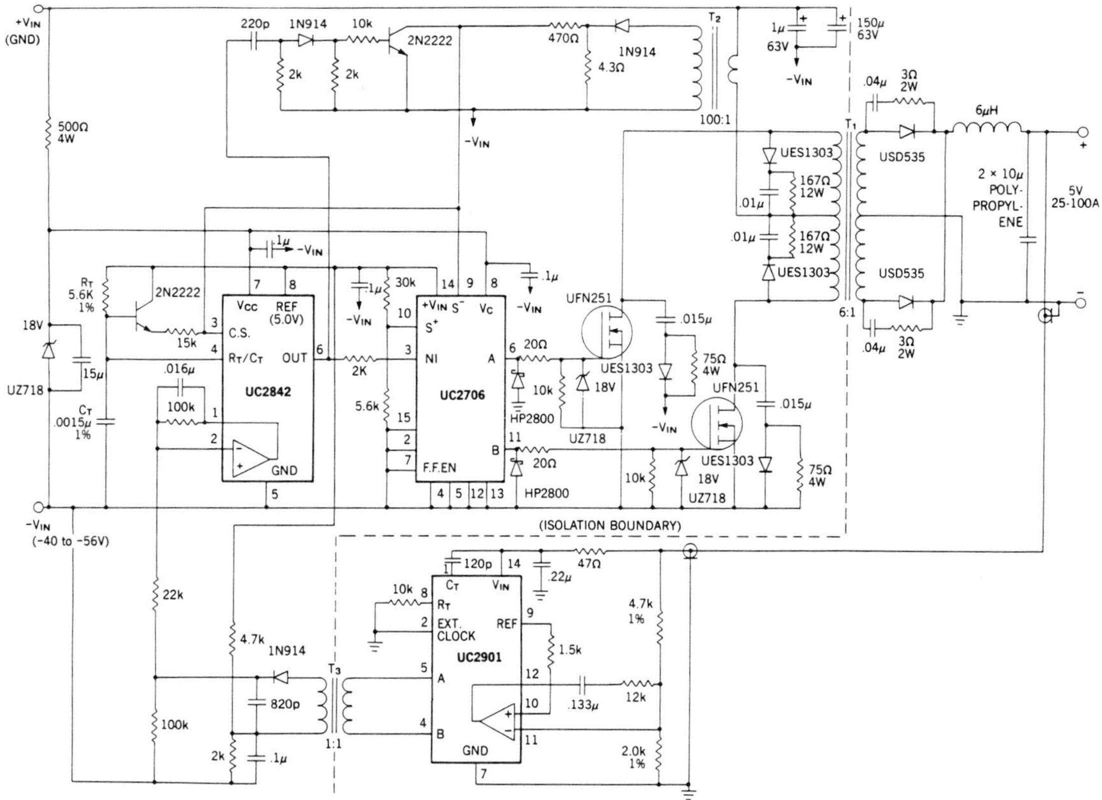

DC-to-DC Push/Pull Converter. This circuit is a 500-W push/pull DC-to-DC converter that uses the UC3842 pulse-width modulation controller, the UC3706 dual output driver, and the UC3901 isolated feedback generator, all from Unitrode. It operates from a standard telecommunications bus to produce a 5-V output at up to 100 A. This design utilizes a center-tap push/pull topology operating with continuous inductor current and current-mode control. This push/pull configuration optimizes transformer utilization while allowing common-source operation of the MOSFETs. Current-mode control was chosen for this application for several reasons. Power transformer flux balancing is achieved without the cost of added sensing circuits. The filter inductor behaves like a current source, which allows a closed-loop frequency response of greater bandwidth than would otherwise be possible for stable operation. More importantly, the error amplifier compensation becomes simpler and better behaved under conditions of large-signal load changes. Finally, current-mode control provides instantaneous (single-cycle) correction for input voltage variations. The UC2842 pulse-width modulator provides all these advantages at an extremely low cost. Since this IC has only a single output, a UC2706 dual-output driver is used to perform a single-ended to push/pull conversion. The UC2706 provides the added advantages of rapid (180 nsec) current limiting and high peak current MOSFET drive needed to operate at high frequencies. Since input-to-output isolation is required, a UC2901 isolated feedback generator is employed to return a feedback signal from the output to the primary-side PWM controller.—"Linear Integrated Circuits Databook." Courtesy of Unitrode Integrated Circuits.

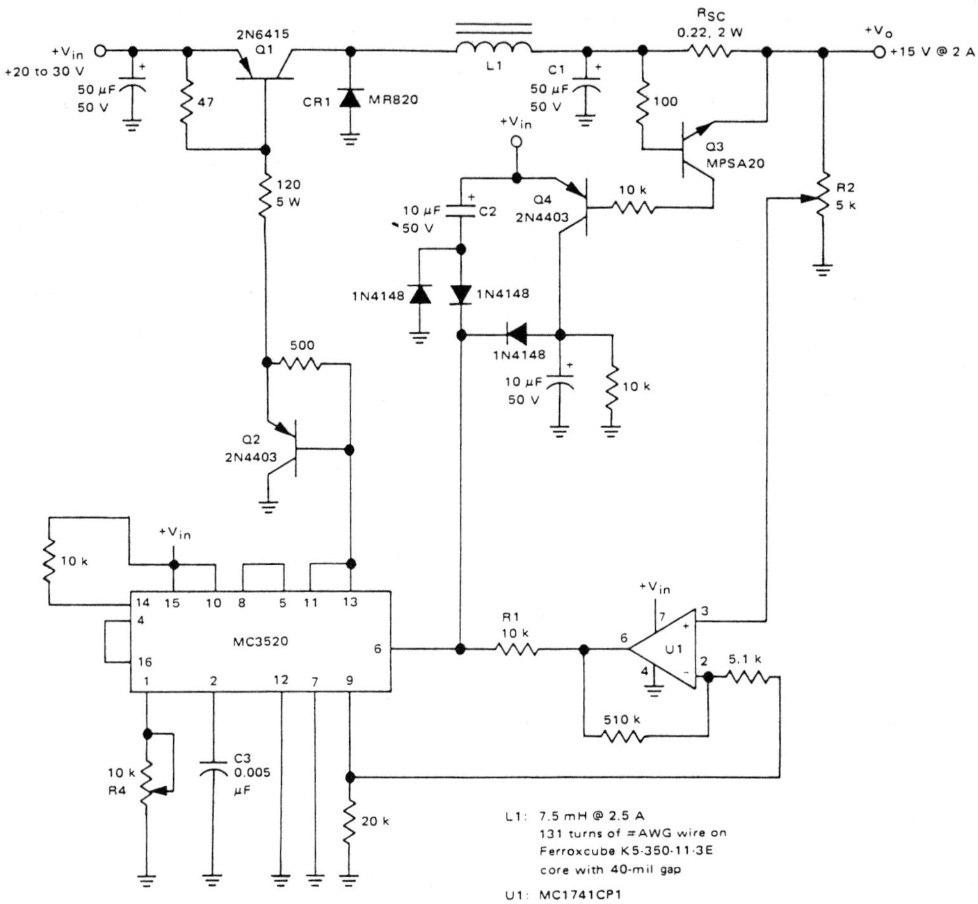

15-Volt DC, 2-Ampere DC-to-DC Converter. This circuit uses the Motorola MC3520 switchmode regulator control IC in a PWM switching power supply. The series switching transistor Q1 chops the input voltage at a frequency of 25 kHz. The resulting waveform is filtered by the LC network consisting of L1 and C1. The frequency is set by C3 and R4. The output voltage is regulated by comparing its value with the reference voltage at the MC3520. C2 provides a soft-start feature during initial power-up. R_{sc}, Q3, and Q4 provide short-circuit protection.—"Linear and Interface Integrated Circuits." Copyright of Motorola, Inc. Used by permission.

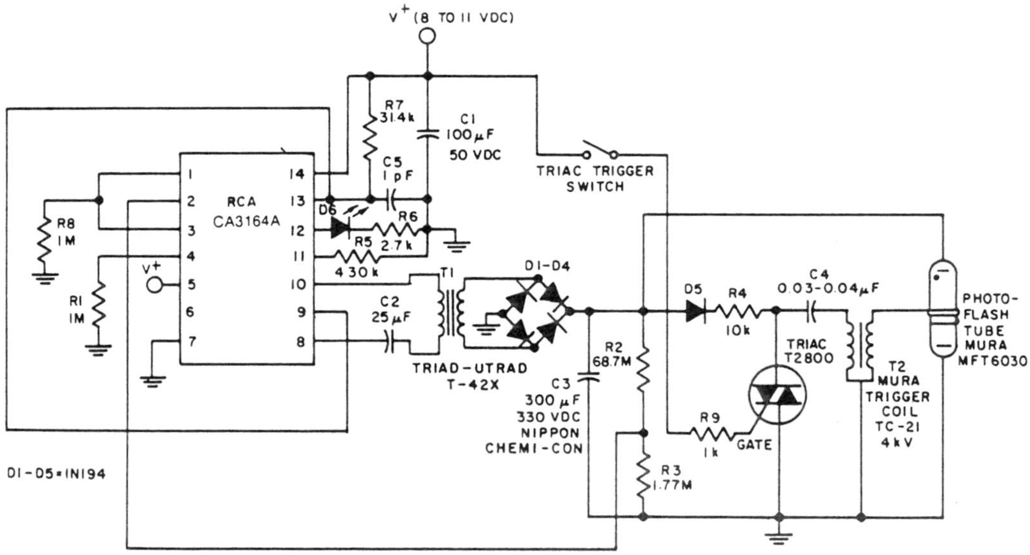

Photoflash Circuit. Low standby current drain helps extend the battery life in portable consumer electronic products that must deliver only occasional bursts of power. A photoflash circuit is ideally representative of this mode of service. The CA3164A control IC is a BIMOS chip that consumes less than 15 μA during standby periods; however, it can still provide 100 mA of chopped current to the DC-to-DC converter in an electronic photoflash during the energy charging cycle. In this circuit, the IC drives the primary of the photoflash step-up transformer (T1) with symmetrically chopped current at a 500- to 2000-Hz rate. Bridge-connected diodes (D1 through D4) rectify the output of T1 and charge the energy storage capacitor C3 to about 280 V. C4 is also charged during this cycle via D5, R4, and the primary of T2. When the camera shutter is released, the triac conducts and C4 is discharged through the primary of T2, generating a short-duration 4-kV pulse across T2's secondary. With the components shown, the IC's chopper frequency is somewhere in the vicinity of 500 Hz.—"RCA Solid State Databook—Integrated Circuits for Linear Applications." Courtesy of Harris Semiconductor.

Chapter 3

Timer Circuits

While this text is written to display the advances that have been made in electronic design due to continued linear IC technology, there is one area which is clearly a throwback to an earlier time. While there are many different types of timer ICs on today's electronics market, the venerable 555 timer is still with us and forms the nucleus for many timing-based circuits used in the most modern of equipment today. However, the 555 is now available in many different forms including dual and quad chips (multiple 555s on a single chip). The almost as venerable 3909 timer chip, better known as an LED flasher, is also still incorporated to a high degree.

Of course, there are other chips and other methods that result in such circuits. The dual comparator is now a mainstay in the field of building a wide range of timing circuits. Special IC types are also available for direct control of TRIACs, and, thus, high-current loads.

All circuits in this section relate to the generic field of "timers." Some use the 555 or its modern day equivalents. Others use different types of linear ICs. No single circuit is so highly specialized that it cannot be incorporated in a myriad of other circuit designs and configurations. While these circuits represent some of the latest technology, some will be little different than those that accomplished the same functions a decade ago.

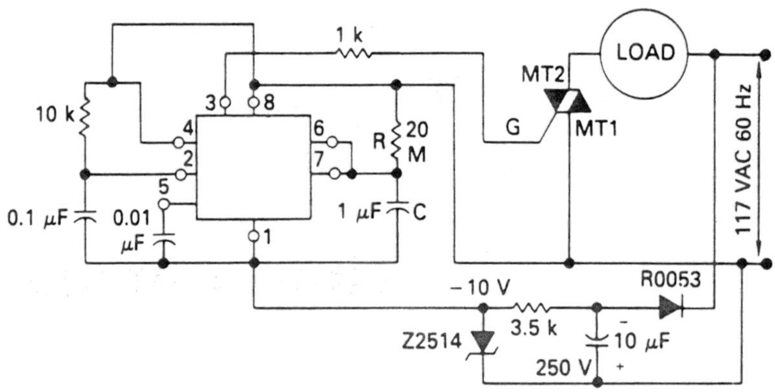

22-Second Solid-State Time Delay Relay Circuit. This circuit makes use of the Motorola MC1455 timing circuit which is a direct replacement for the generic 555 timer that is available from every IC manufacturer. Here, the timer is used as the trigger for the TRIAC which, in turn, controls the load. Note that the 22-second delay can be varied by changing the values of R and C. Backing off on the value of R and/or decreasing the value of C will produce shorter time delays. Longer delays can be produced by increasing the values of these components.—"Linear and Interface Integrated Circuits." Copyright of Motorola, Inc. Used by permission.

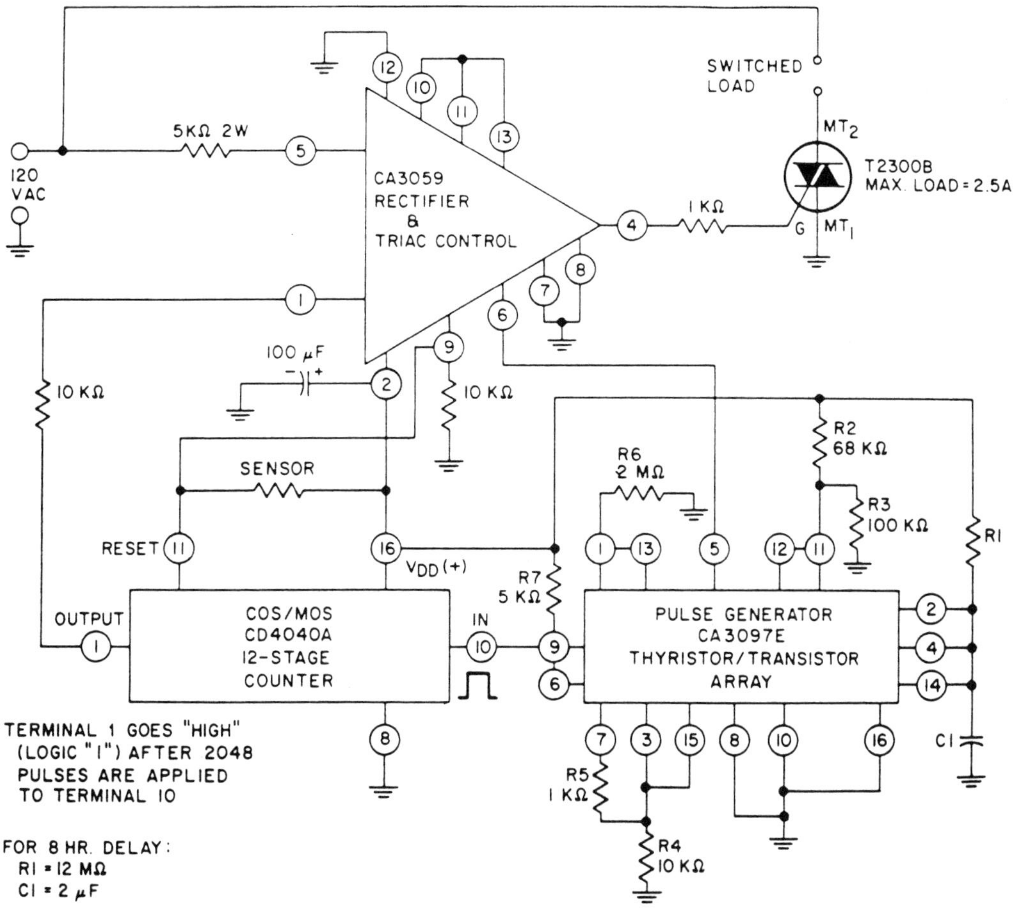

Eight-Hour Timer. This is an extremely long duration, line-operated IC timer. It utilizes the RCA CA3059 rectifier and TRIAC control to drive a T2300B TRIAC with a maximum throughput of 2.5 A. A 12-stage counter IC and the CA3097E pulse generator/array are also incorporated. R1 and C1 control the delay timing. For an 8-hour delay, R1 should be a 12-MΩ unit, while C1 must have a value of 2 μF. If higher current is desired, the small TRIAC in this circuit may be used to control a relay, which in turn switches a much higher power level. Also, this TRIAC can be replaced with an SCR, which can be used to drive a higher-powered TRIAC.—"RCA Solid State Databook—Integrated Circuits for Linear Applications." Courtesy of Harris Semiconductor.

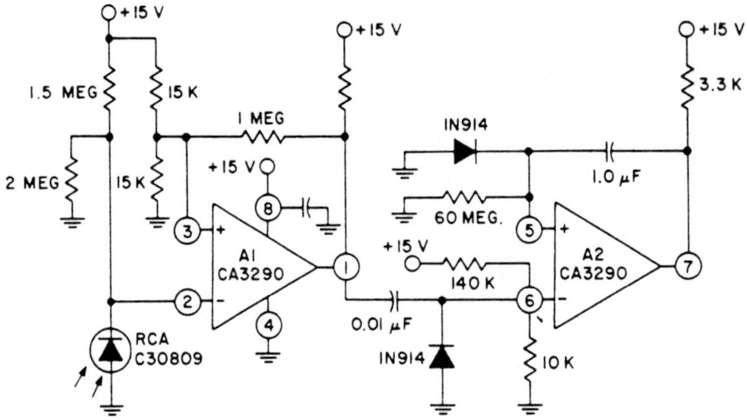

Light-Controlled One-Shot Timer. The CA3290 IC is a dual comparator chip and one comparator is used to sense a change in photo diode current. The other is configured as a one-shot timer and is triggered by the output of the first comparator. The output of the circuit will switch to a low state for approximately 60 sec after the light source to the photo diode has been interrupted. The circuit is designed to operate at normal room lighting levels.—"RCA Solid State Databook—Integrated Circuits for Linear Applications." Courtesy of Harris Semiconductor.

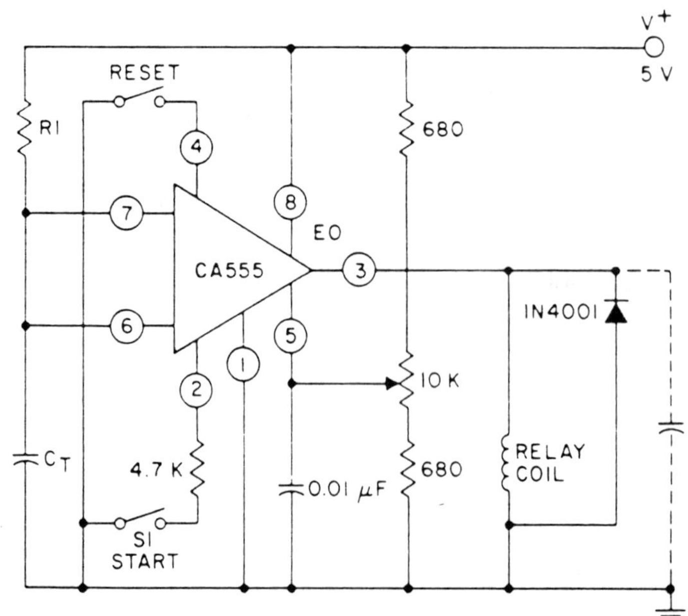

Repeat Cycle Timer. Similar to the previous circuit, this example is connected as a repeat cycle timer where the total period is a function of R1 and R2.—"RCA Solid State Databook—Integrated Circuits for Linear Applications." Courtesy of Harris Semiconductor.

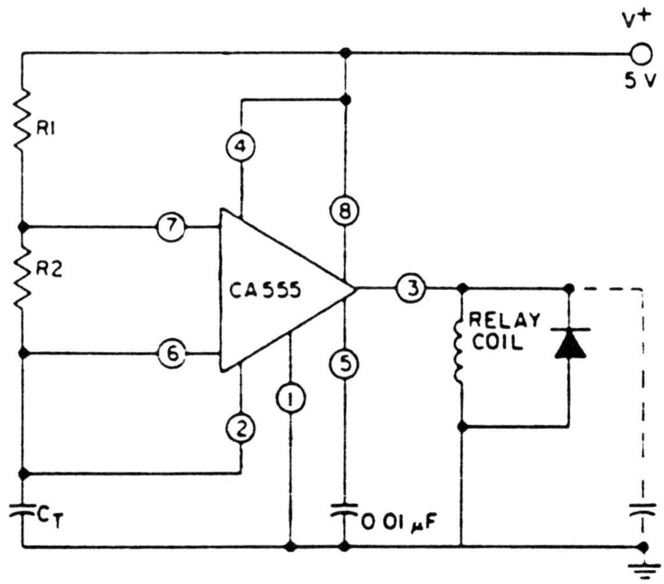

◀ **Reset Timer.** Using the 555 IC timer (RCA designation CA555) in a monostable operation, this circuit holds C_t in an initially discharged state via a transistor within the IC. Upon closing the "start" switch (or applying a negative trigger pulse to terminal 2), the integral timer flip-flop is set and releases the short circuit across C_t which drives the output voltage to a high state, activating the relay. The voltage across C_t increases to a point equal to approximately two-thirds of the supply voltage. At this point, the comparator resets the flip-flop which discharges the capacitor rapidly and drives the output to a low state. This circuit is easily miniaturized. A small relay can be mounted directly on the circuit board or it can be located at a remote point.—"RCA Solid State Databook—Integrated Circuits for Linear Applications." Courtesy of Harris Semiconductor.

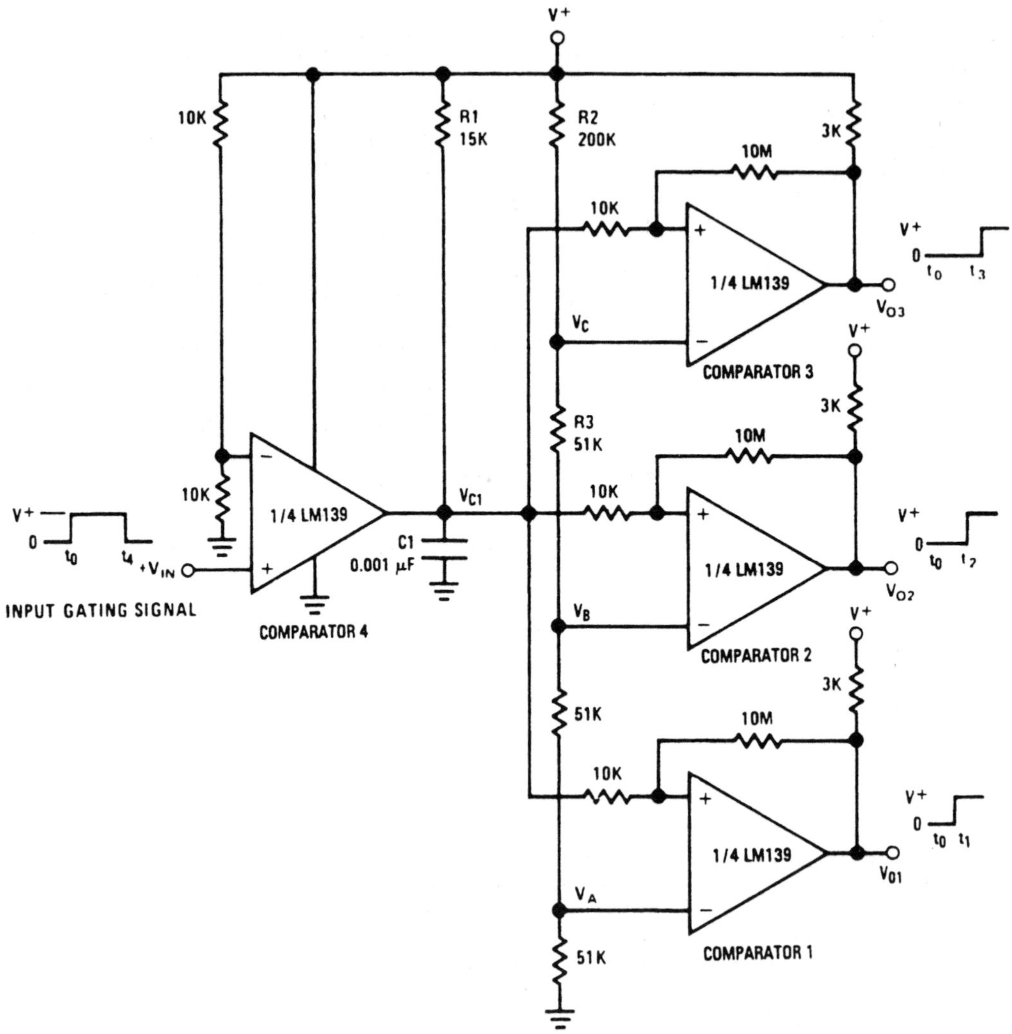

Time Delay Generator. This timer will provide output signals at prescribed time intervals from a time reference (T_0) and will automatically reset when the input signal returns to ground. In the quiescent state, the output of comparator 4 is *on*, which keeps the voltage across C1 at zero. This keeps the outputs of the other comparators in their *on* states. When an input signal is applied, comparator 4 turns *off* allowing C1 to charge through R1. As this voltage rises the output voltage of each of the remaining comparators will switch to the high state. A small amount of hysteresis is provided to facilitate fast switching times. The National Semiconductor LM139 is an IC that contains four independently functioning comparators. This chip may be powered from a single-polarity supply that offers a very wide range of from 2 to 36 V DC.—"Linear Applications Databook." Reprinted with permission of National Semiconductor.

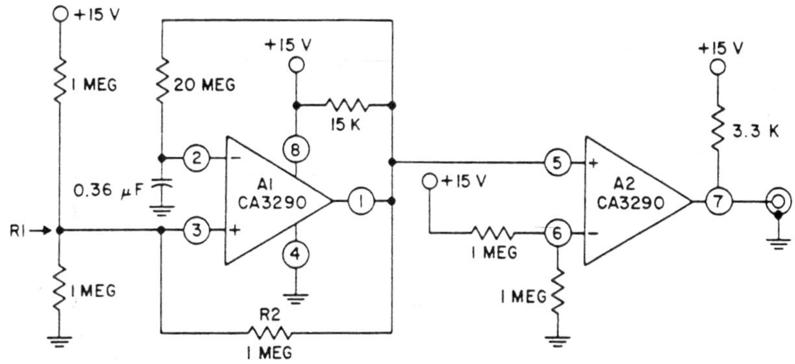

Low-Frequency Multivibrator. This multivibrator uses the CA3290 dual voltage comparator IC. The first comparator is a conventional multivibrator configuration. Owing to the extremely high input impedance of this device, large values of the timing resistor (R1) may be used for long time delays with relatively small leakage timing capacitors. The second comparator in the chip serves as an output buffer to ensure that the output frequency from the first comparator is not "pulled" by output loading.—"RCA Solid State Databook—Integrated Circuits for Linear Applications." Courtesy of Harris Semiconductor.

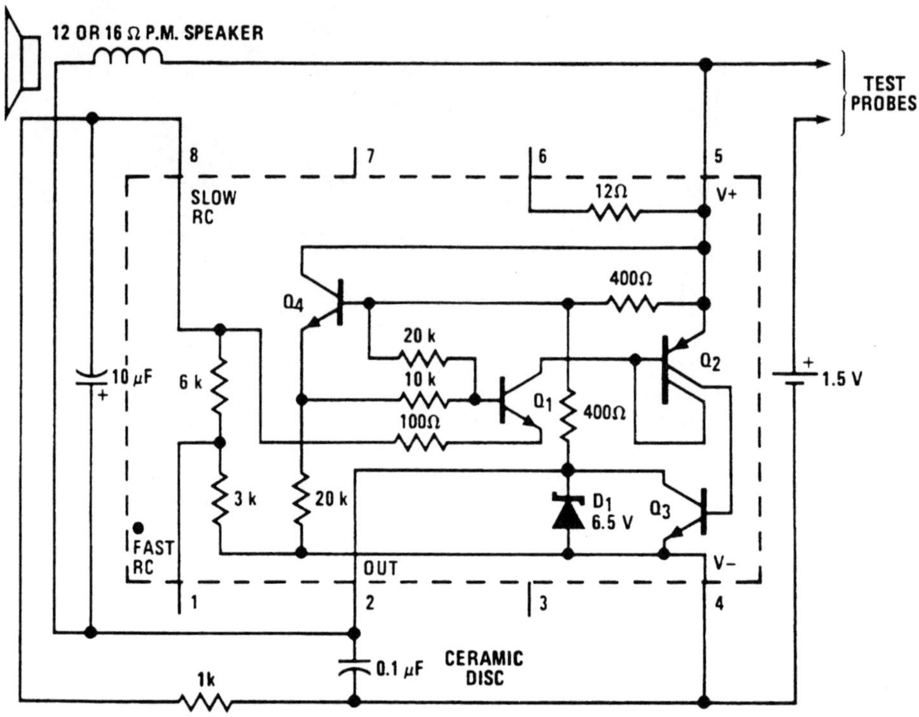

Continuity Checker. This is an audible continuity checker that indicates continuous circuits by sound rather than by the position of a needle on a meter. The heart of the circuit is the very versatile LM3909 LED flasher/oscillator. Basically, this circuit is a "dirty" oscillator which is triggered whenever the resistance between the test probes is less than approximately 100 Ω. External components consist of only a 10-μF electrolytic capacitor, a 1-kΩ resistor, a 0.1-μF disk ceramic capacitor, and a PM speaker. Power is provided by a single 1.5-V battery which may be of the AA or even AAA type for the smallest package size possible. No transformer is necessary between the IC output and the speaker because the 150-mA capability of the output is adequate to drive small speakers directly.—"Linear Applications Databook." Reprinted with permission of National Semiconductor.

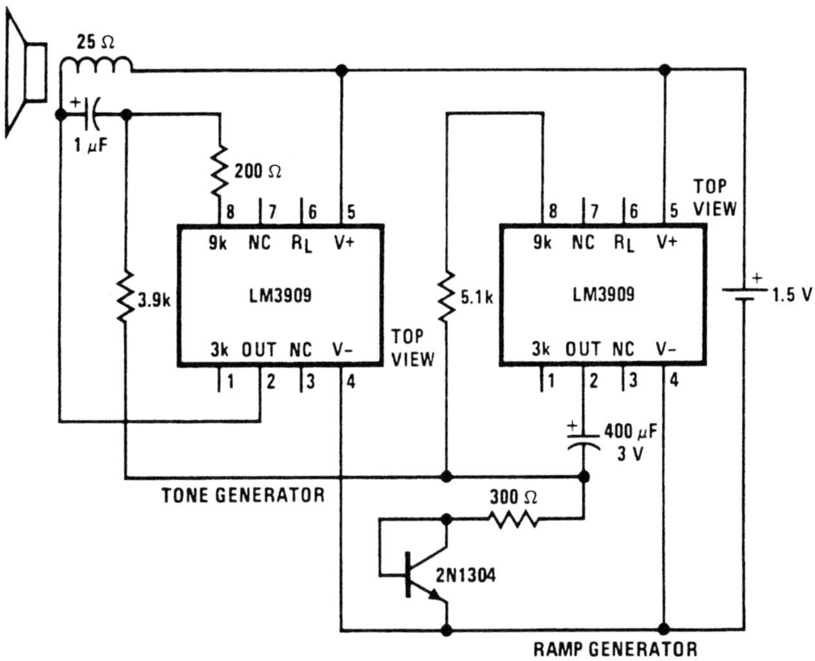

Whooper Siren. By combining two LM3909 flasher ICs a very effective whooper siren effect can be produced. The first IC acts as a tone generator while the second is the ramp generator, which produces a rapidly rising and falling modulating voltage in conjunction with the 400-μF capacitor. Notice that the transistor is short-circuited between its base and collector. This means that it really looks like a low-voltage germanium diode to the circuit. This device, along with the 5.1-kΩ feedback resistor forces the ramp generator into an unusual mode of operation having longer *on* states than *off* states. This raises the average frequency of the tone generator and causes the modulation to seem more even. Try playing with the values of the electrolytic capacitor and the feedback resistor for more interesting sound effects.—"Linear Applications Databook." Reprinted with permission of National Semiconductor.

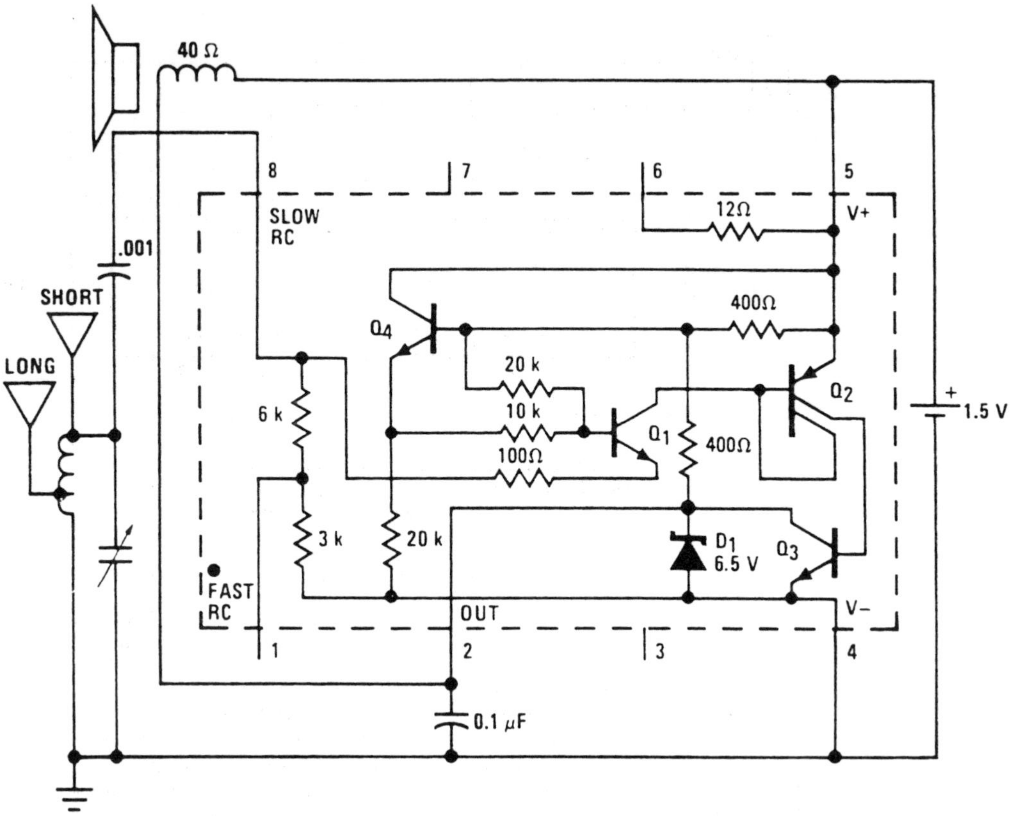

Flasher IC Used as an AM Radio. This circuit is a bit strange and practically useless, being no better than a simple crystal set radio receiver. It does make an interesting practicum, however. The IC is the LM3909 LED flasher/oscillator, but in this mode it neither flashes nor oscillates. Instead it acts as a detector amplifier. It does not oscillate because there is no feedback path from pin 2 to pin 8. Local radio stations can be detected by this simple arrangement that is powered by a single 1.5-V DC battery. The antenna coil is a standard ferrite rod antenna that should be available from any hobby store. Short antenna wires on the order of 20 ft or less work best when attached to the top of the coil. Longer antennas provide better sensitivity when connected to the coil tap which is nearer the bottom; however, different locations will yield different results, so experiment here.—"Linear Applications Databook." Reprinted with permission of National Semiconductor.

Chapter 4

Oscillator Circuits

Those readers who can remember the earlier years of electronics when the vacuum tube reigned supreme are quite knowledgeable of the strenuous (by today's standards) physical design requirements of even a simple 100-kHz oscillator circuit. Even though it was usually a single-tube design, this component required a multivoltage power supply, tube socket, chassis, and so on. The idea of "throwing" together a simple oscillator in a half hour or so was out of the question. Then came the transistor and things got simpler.

However, the linear integrated circuit has ingratiated itself to every area of electronics design and especially to the field of high-frequency output devices. As a matter of fact, a single linear IC type may be made to serve in an oscillator or multivibrator configuration that outputs frequencies from a few hertz to several hundred megahertz. And all of this is a single package, requiring one low-voltage, low-current power supply with the entire works being "glued" to a single scrap of circuit board.

The circuits in this chapter can be classified as belonging to the category of oscillators. Their uses run the gamut from test circuits to stable drivers for complex communications equipment.

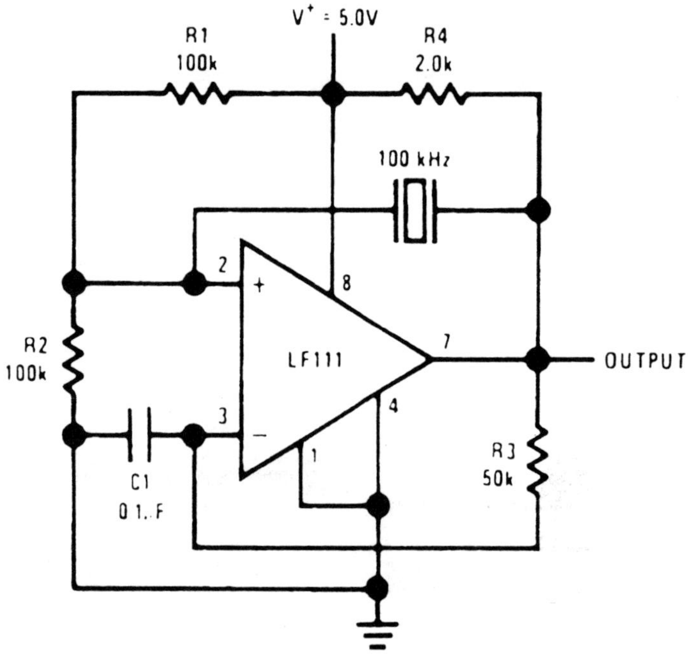

100-Kilohertz Crystal Oscillator. Using the LF111 voltage comparator, this compact circuit requires only a few components to produce a stable 100-kHz output. While configured for a 5-V DC supply, this IC is capable of operation at potentials as high as 36 V DC.—"Linear Databook I." Reprinted with permission of National Semiconductor.

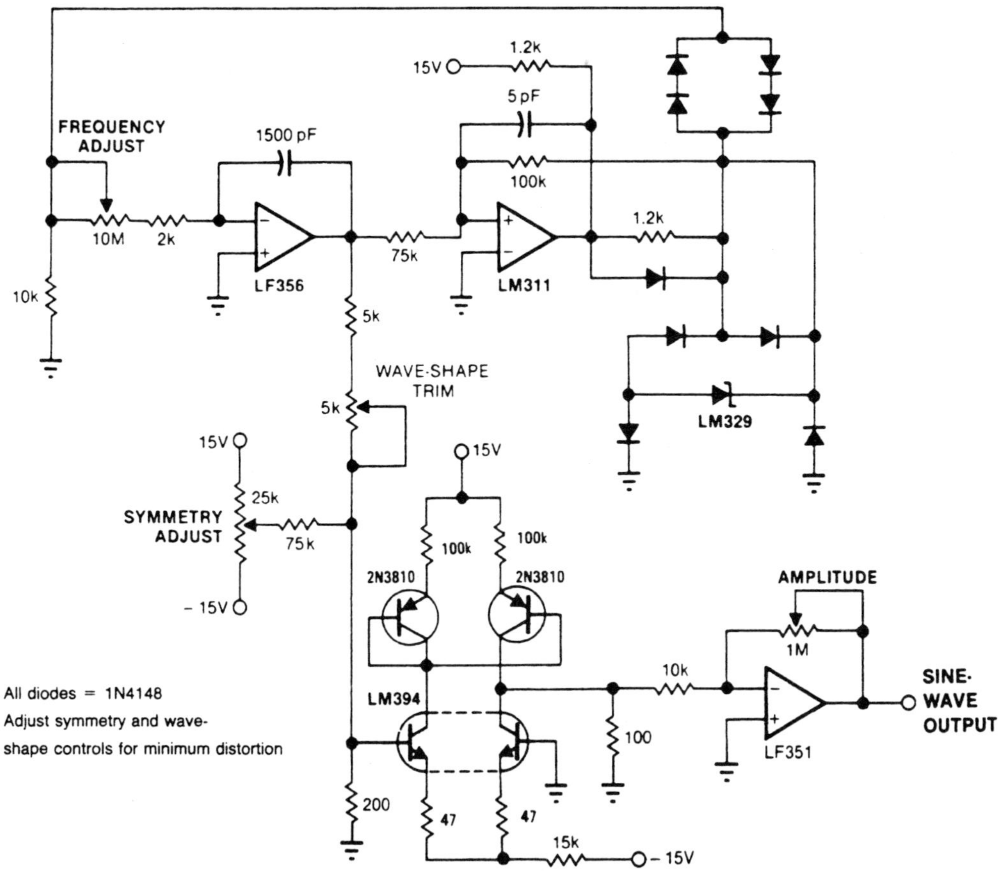

1-Hertz to 10-Kilohertz Sine Wave Oscillator. This is an extremely broad-ranged sine wave oscillator that uses logarithmic shaping schemes. Amplitude stability and distortion characteristics are both excellent. The circuit works by placing an integrator inside the positive feedback loop of the LM311 comparator. This IC drives a symmetrical, temperature-compensated clamp arrangement. The output of the clamp biases the LF356 integrator. This IC integrates the current into a linear ramp at its output. This ramp is summed with the clamp output at the LM311 input. When the ramp voltage nulls out the bound voltage, the comparator changes state and integrator output is reversed. The LM394 dual transistor IC is used to generate sine shaping while the 2N3810 provides current drive. The LF351 allows adjustable, low-impedance, output amplitude control.—"Linear Applications Databook." Reprinted with permission of National Semiconductor.

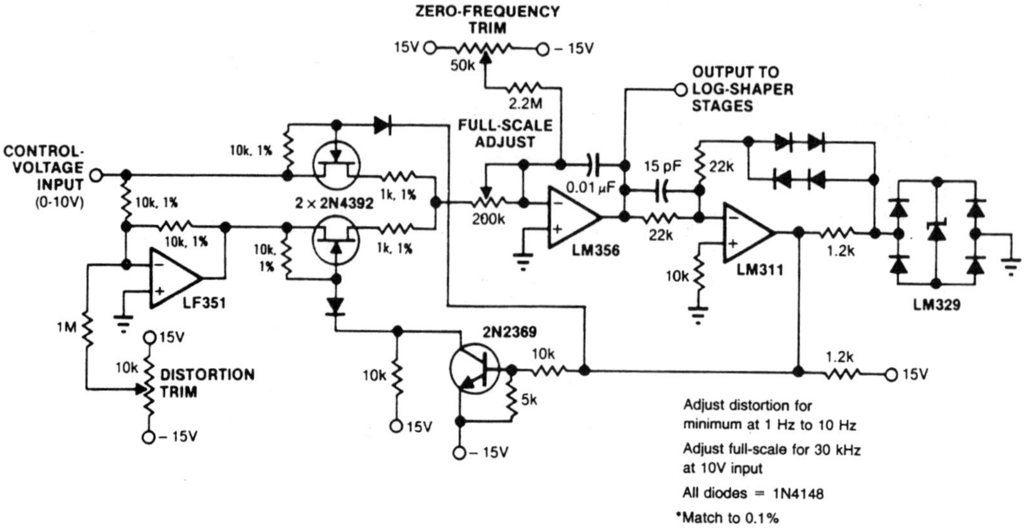

Improved Sine Wave Oscillator. This is a modification of the previous circuit that includes signal-level-controlled feedback. Here, FETs switch the integrator's input so that the resulting summing-junction current is a function of the input control voltage. The configuration results in a frequency range of 1 Hz to 30 kHz for a 0- to 10-V input.—"Linear Applications Databook." Reprinted with permission of National Semiconductor.

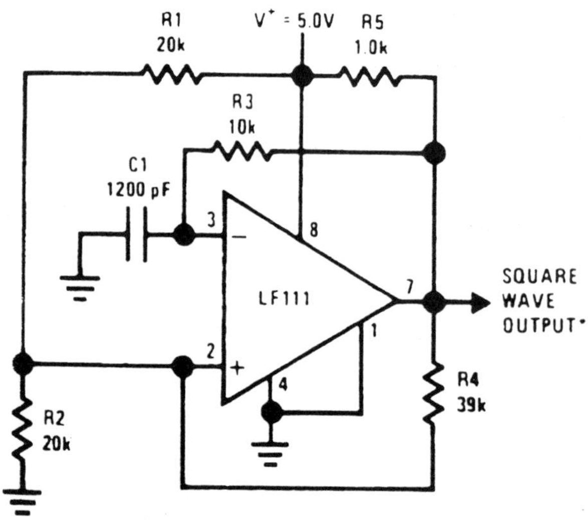

100-Kilohertz Free-Running Multivibrator. The LF111 voltage comparator IC is used in this configuration to produce a 100-kHz square wave output. The circuit operates from a single-polarity 5-V DC source and is extremely compact, requiring only five resistors and a single capacitor as external components.—"Linear Databook I." Reprinted with permission of National Semiconductor.

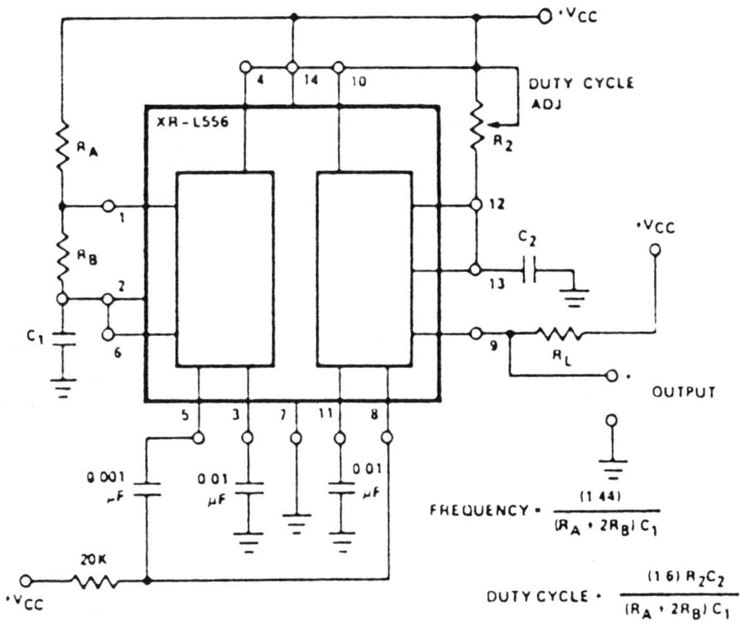

Micropower Oscillator. This oscillator circuit uses the dual 555 timer chip (XR-L556) from EXAR and offers fixed-frequency operation along with a variable duty cycle. The chip contains two independent 555 timers, the first operating in astable mode and the second in monostable mode. This results in an oscillator with a fixed frequency and a variable duty cycle. One timer generates a basic period waveform that is then used to trigger the second timer. The time delay of this second timer element is less than the period of oscillation of the first timer, then the output at pin 9 has the same frequency as the first timer. But, its duty cycle is determined by the timing cycle of the second timer. Using this circuit with components shown, the output duty cycle can be adjusted over a very wide range from 1 to 99%. The duty cycle control R2 is a linear taper potentiometer.—"EXAR Databook." Reproduced with the express authorization of EXAR Corporation.

Cheap and Simple 16-Megahertz Oscillator. In addition to the Signetics NE5212 transimpedance amplifier IC, all that is needed to build this oscillator is a 2-pF capacitor and a crystal. This circuit operates in the series resonant mode. The noninverting input is fed back to the input of the IC in series with the 2-pF capacitor. The oscillator output is taken from the inverting output.—"Linear Data Manual Volume 2: Industrial." Courtesy of Signetics Company, a division of North American Philips Corporation.

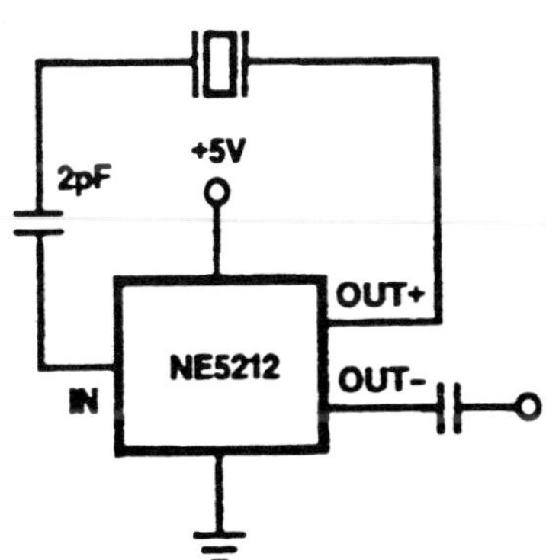

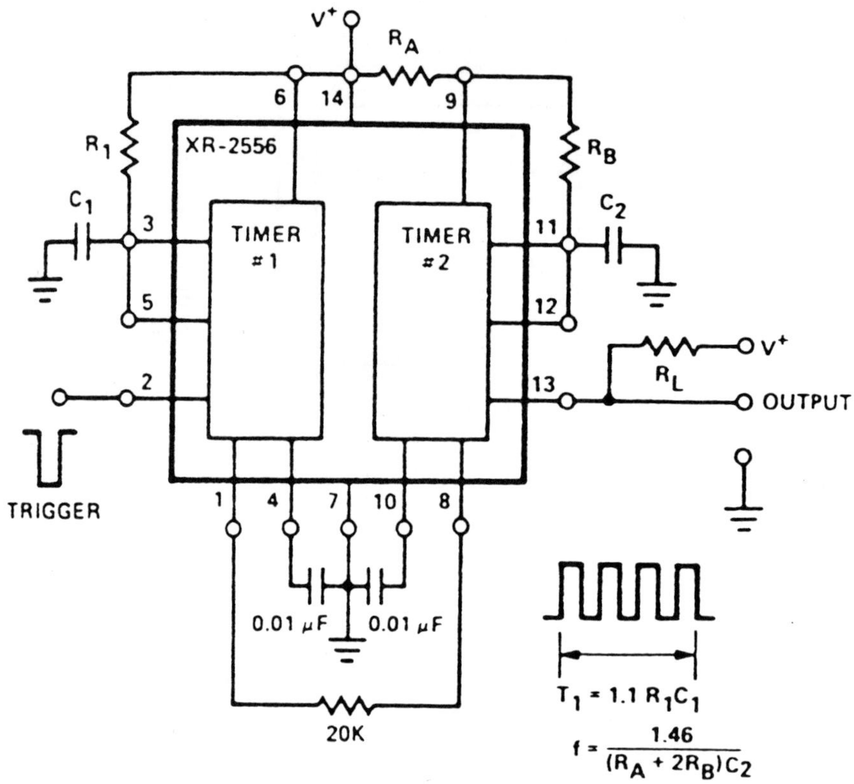

Keyed Oscillator. The XR-2556 from EXAR contains two independent 555 timers on a single chip. In this configuration, one of the timers is operated in its free-running mode, while the other timer section is used to key the first on and off. In this schematic, timer 2 is the oscillator section. Its frequency is set by the values of R_a, R_b, and C2. Timer 1 is operated as a monostable circuit and its output is connected to the reset terminal (pin 8) of timer 2. When the circuit is at rest, the logic level at the output of timer 1 is low and the oscillations of timer 2 are inhibited. Upon application of a trigger signal to the first timer, the logic level at pin 1 goes high and the second timer is keyed on. The output of timer 2 appears as a tone burst whose frequency is set by the aforementioned components and whose duration is set by R1 and C1.—"EXAR Databook." Reproduced with the express authorization of EXAR Corporation.

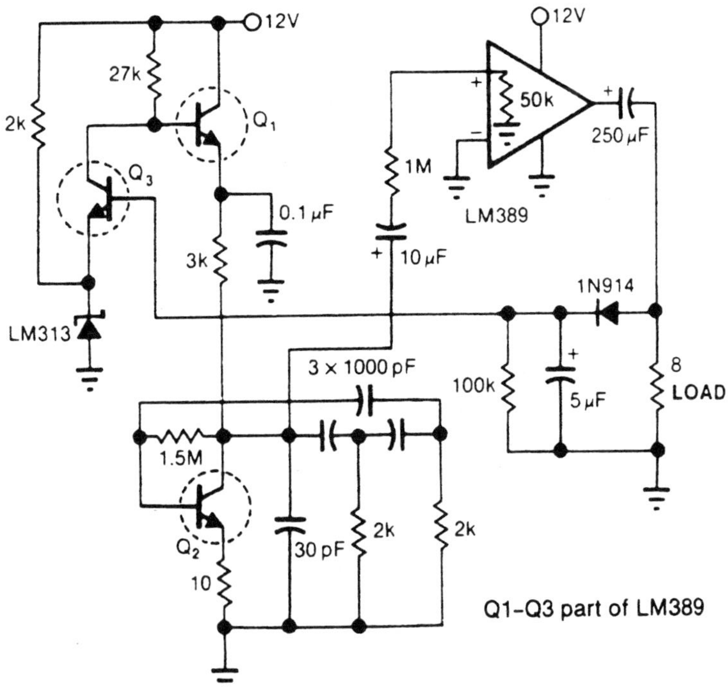

Phase-Shift Oscillator. This is an inexpensive amplitude stabilized phase-shift sine wave oscillator circuit that requires only a single IC package and operates from a single-polarity 12-V DC supply. The LM389 audio power amplifier IC contains three discrete N-P-N transistors in addition to the amplifier proper. Q2, in combination with the RC network, comprises a phase-shift configuration and oscillates at about 12 kHz. The remaining circuitry provides the amplitude stability. Note that the three transistors shown in this schematic are actually a part of the IC package and are illustrated discretely for clarity of understanding. The high-impedance output at the collector of Q2 is fed to the input of the power amplifier through a 10-μF capacitor and 1-MΩ carbon resistor series RC network. This effectively divides the output from Q2 by a factor of 20, which is necessary because the power amplifier has a fixed gain of 20. Therefore, the amplifier functions as a unity-gain current buffer that drives an 8-Ω load.—"Linear Applications Databook." Reprinted with permission of National Semiconductor.

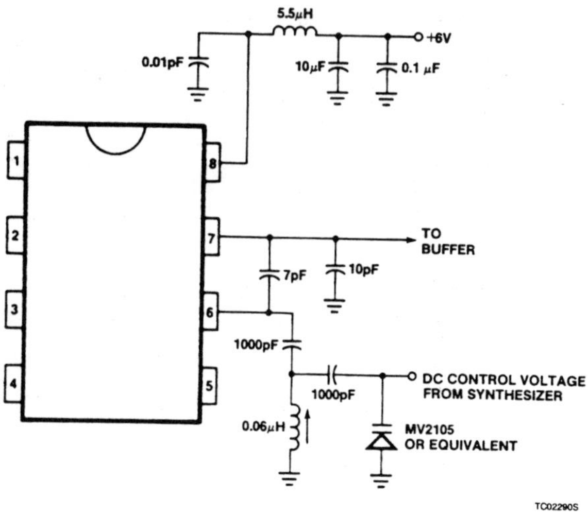

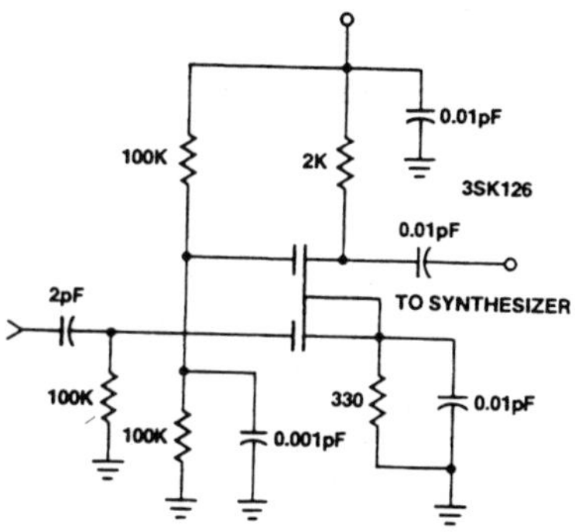

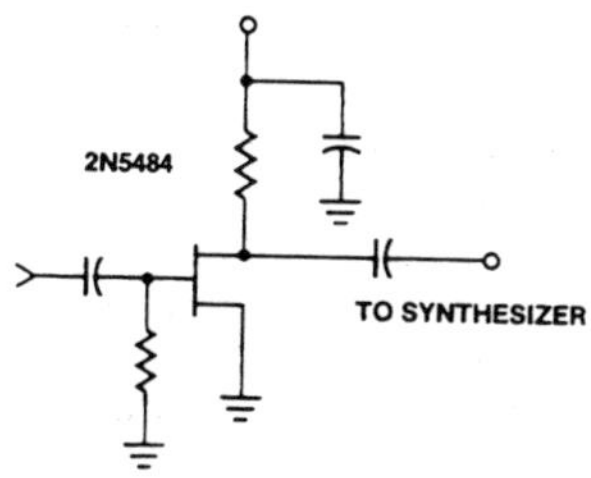

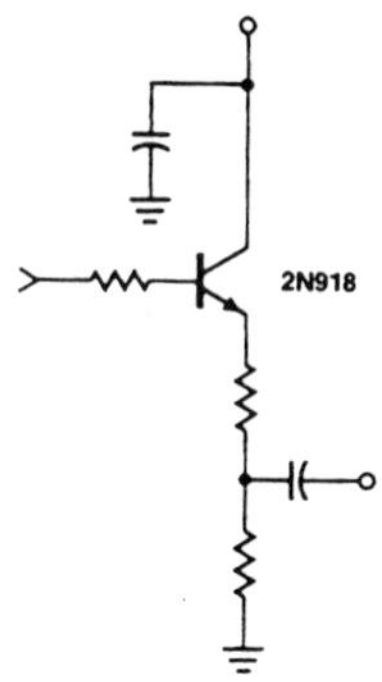

Colpitts Oscillator for Synthesizer Applications. The Colpitts varactor-tuned tank oscillator shown in this schematic has immediate applications in synthesizer-controlled environments. It is important to buffer the output of the circuit in order to assure that switching spikes from the first counter or prescaler do not end up in the oscillator spectrum. This circuit uses the Signetics NE612 double-balanced mixer and oscillator IC.—"Linear Data Manual Volume 1: Communications." Courtesy of Signetics Company, a division of North American Philips Corporation.

Crystal Oscillator. This circuit uses the RCA 311 comparator IC as a fundamental crystal oscillator at 100 kHz. Driven from a 5-V DC power supply, it requires a minimum of external components. The quartz crystal may be paralleled by a small value variable capacitor (piston trimmer type preferred) to allow for a small variation in output frequency.—"RCA Solid State Databook—Integrated Circuits for Linear Applications." Courtesy of Harris Semiconductor.

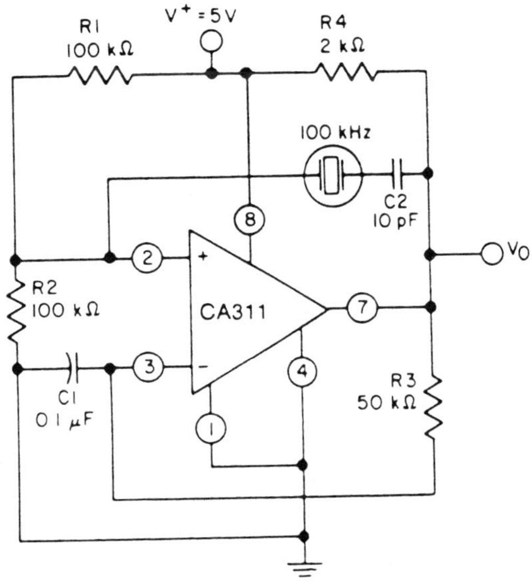

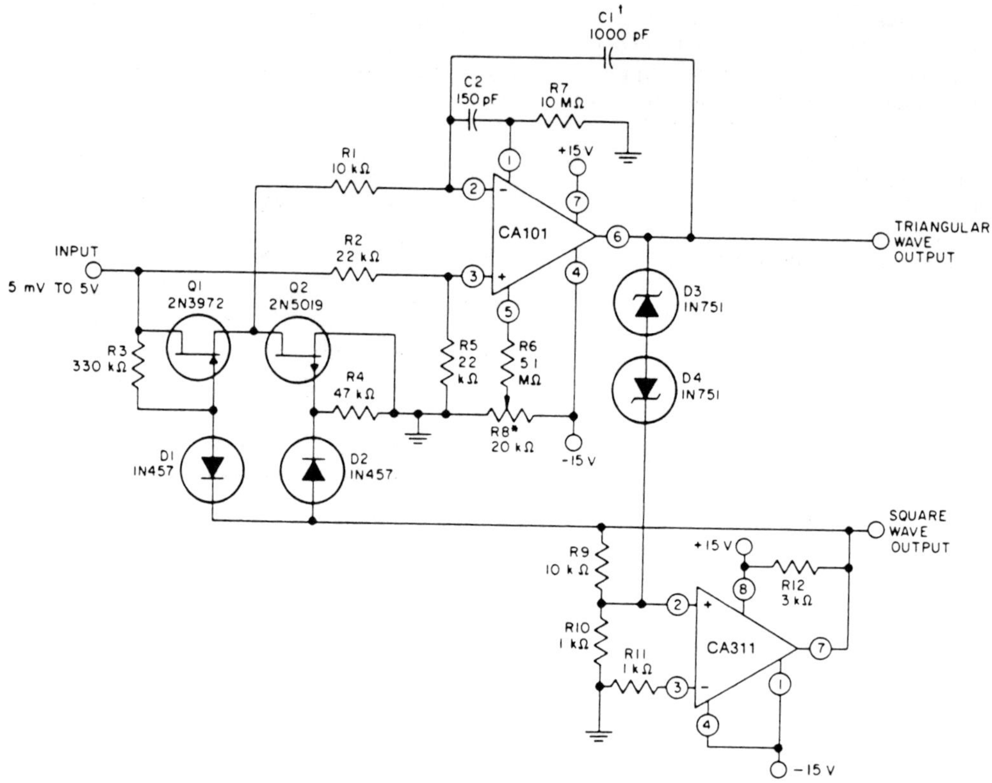

10-Hertz to 10-Kilohertz Voltage-Controlled Oscillator. Utilitizing the RCA 311 Voltage comparator at the output of a 101 op amp, this circuit produces a square wave output. R8 is used to adjust for a symmetrical square wave output time when V_i is equal to 5 mV.—"RCA Solid State Databook—Integrated Circuits for Linear Applications." Courtesy of Harris Semiconductor.

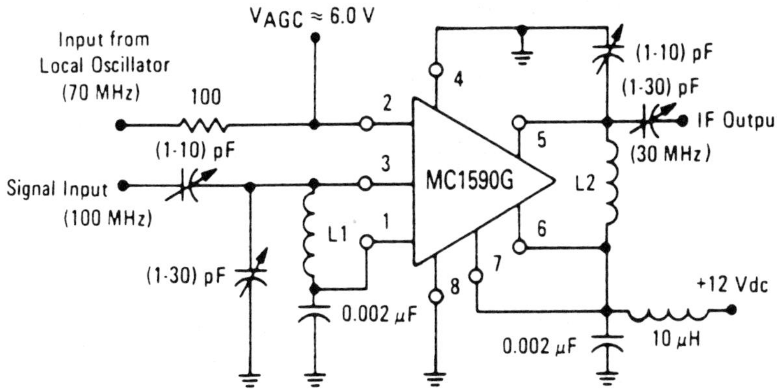

100-Megahertz Mixer. Using the MC1590G (Motorola) wideband amplifier IC, this circuit derives its power from a single 12-V DC source and makes provisions for AGC. For the frequencies specified, L1 consists of 5 turns of #16 AWG wire $\frac{1}{4}$ in. in diameter and spaced evenly for a length of $\frac{5}{8}$ in. L2 is 16 turns of #20 AWG wire on a toroid core (t44-6 Micro Metal or equivalent).—"Linear and Interface Integrated Circuits." Copyright of Motorola, Inc. Used by permission.

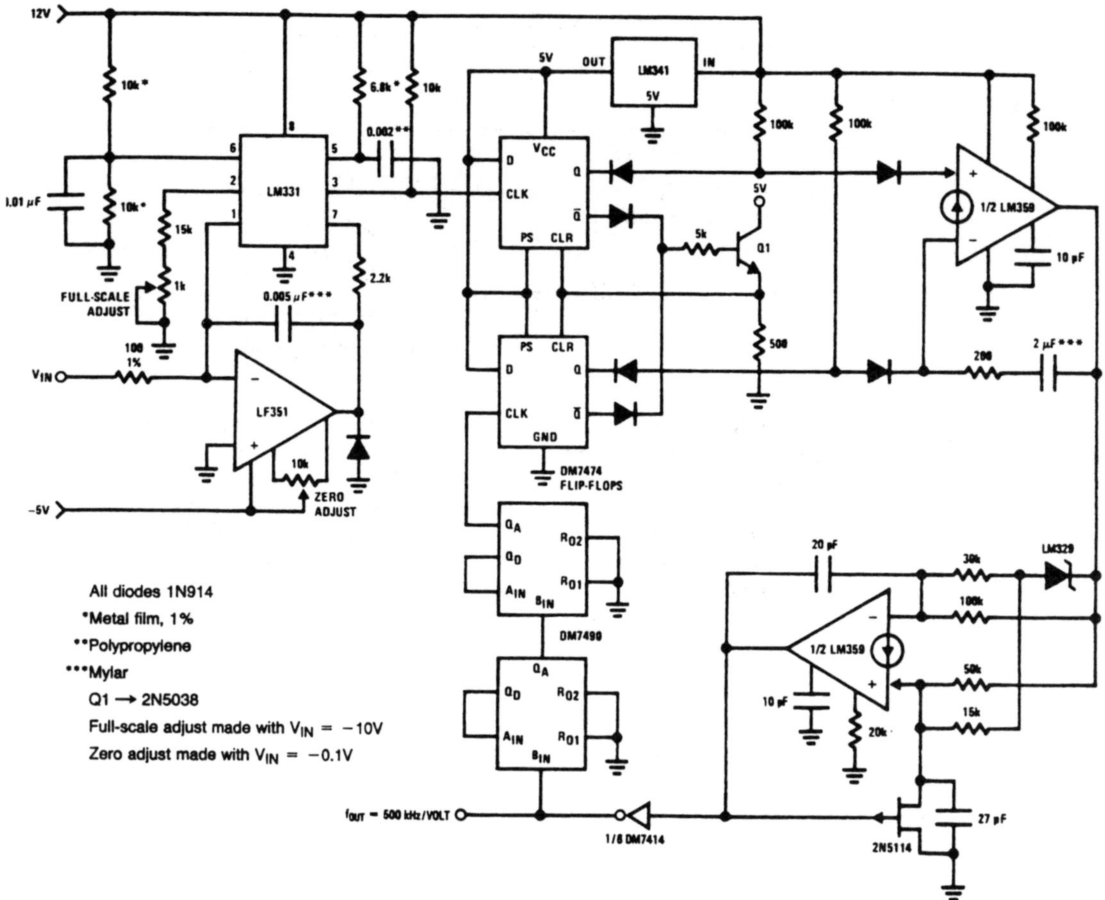

Ultralinear, Two-Decade VCO. Linear and fast voltage-to-frequency (V/F) converters are extremely arduous to build, especially when standard V/F design techniques are used. One way around this difficulty is to use a fast phase-locked loop driven by a medium-frequency and ultralinear V/F IC. In this case, the LM331 V/F IC is incorporated. The high-frequency operation is obtained by means of a frequency divider inserted into the loop. The linearity of the overall circuit is a close approximation of the linearity of a medium-frequency V/F. The high-frequency, quasilinear VCO and the error amplifier of the PLL are designed by using the two sections of an LM359 Norton amplifier. The output frequency of the VCO, which is also the output of the system, is divided by 100 and is compared with the output of the driving V/F via a digital phase detector. Following a zero and a full-scale adjust this circuit works well over two decades of frequency (50 kHz to 5 MHz) and its nonlinearity is below 0.03%.—"Linear Applications Databook." Reprinted with permission of National Semiconductor.

Chapter 5

Computer and Digitally Oriented Circuits

While the emphasis on integrated circuits has been directed more at the digital side of the coin, the fact is that the interfacing of digital and linear integrated circuits is resulting in some very efficient, useful, and powerful gadgetry. This chapter features designs that either incorporate digital devices with linear ICs or use linear ICs entirely to address certain computer-related applications.

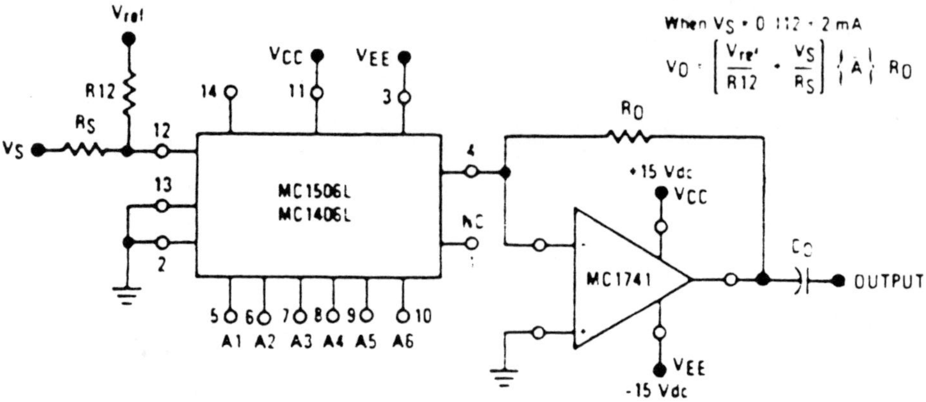

Digital Attenuator. The Motorola MC1506L DAC is used in the multiplying mode to form a digital attenuator. One advantage of this technique is that if $R_s = 50\ \Omega$, no compensation capacitor is needed and a wide, large signal bandwidth is achieved.—"Linear and Interface Integrated Circuits." Copyright of Motorola, Inc. Used by permission.

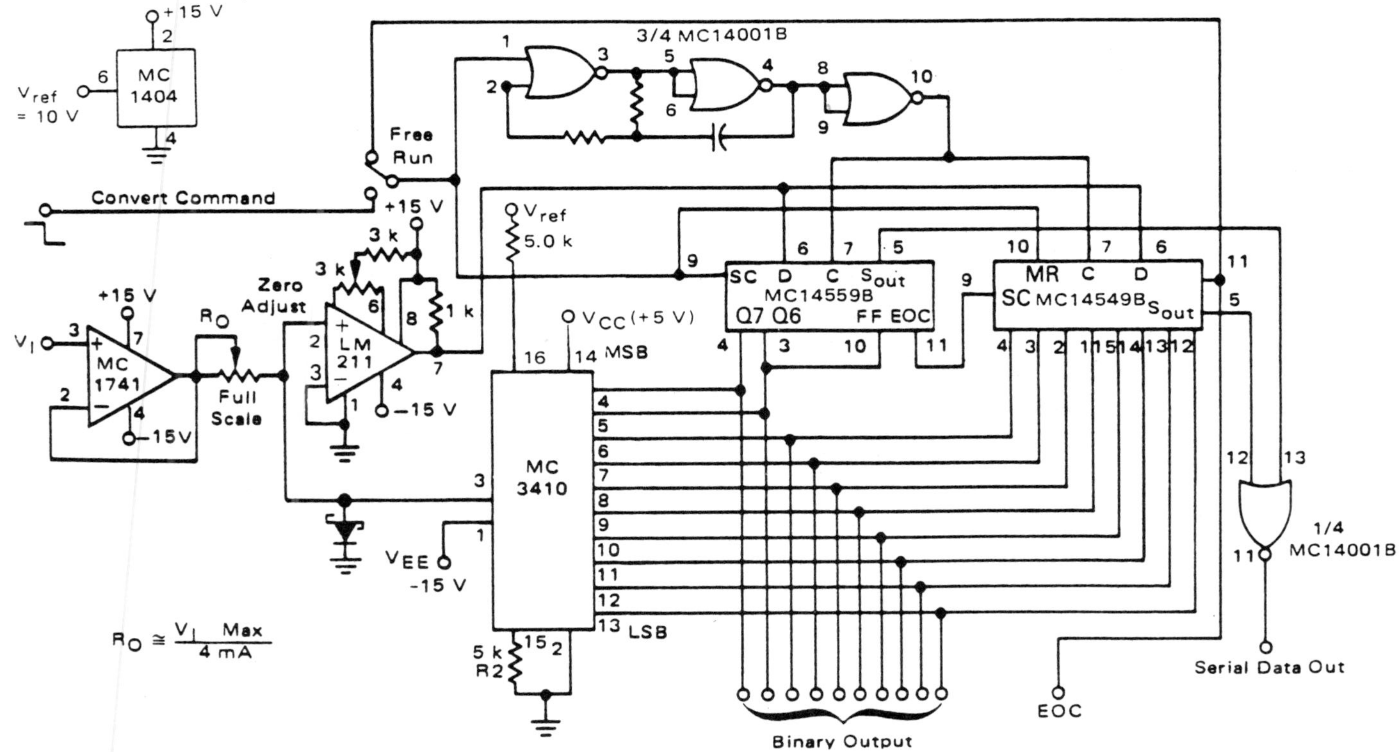

Successive Approximation ADC. The fastest, most efficient means of analog-to-digital conversion using DACs is successive approximation. Similar in appearance to staircase devices, the successive approximation (SA) converter is capable of 100 times faster conversions for a 10-bit result. The complete SA converter is shown in this schematic using the Motorola MC3410 and MC14559B/49B successive approximation registers. The complexity that results in the faster conversion speed is found in the latter's registers. The register compares the DAC output resulting from activating each bit with the input voltage. This is done starting with the most significant bit and after 10 comparisons generates the 10-bit binary output representing the input. The accuracy of this conversion is determined by the MC3410 and is not dependent on tolerances of the other components.—"Linear and Interface Integrated Circuits." Copyright of Motorola, Inc. Used by permission.

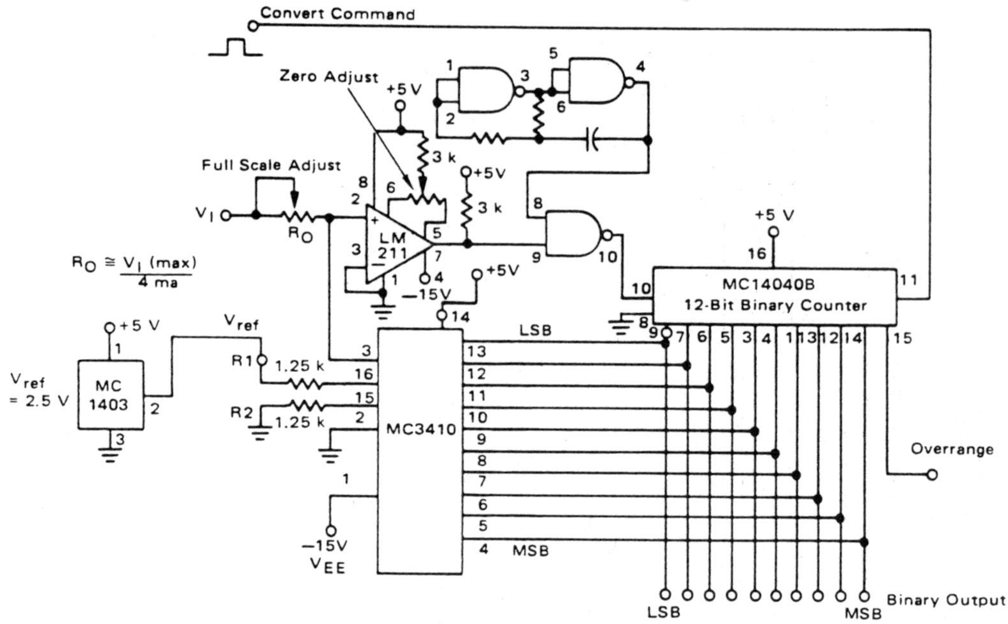

Staircase Analog-to-Digital Converter. This schematic shows a complete staircase A/D converter that uses several Motorola chips. With an input voltage applied, the binary counter is reset by the convert command pulse and the "begin accumulating" counts. The DAC output steps upward until the comparator detects that the input is equal to the DAC output. The counters are disabled and the conversion result is held at the output until the circuit is reset by the convert command input.—"Linear and Interface Integrated Circuits." Copyright of Motorola, Inc. Used by permission.

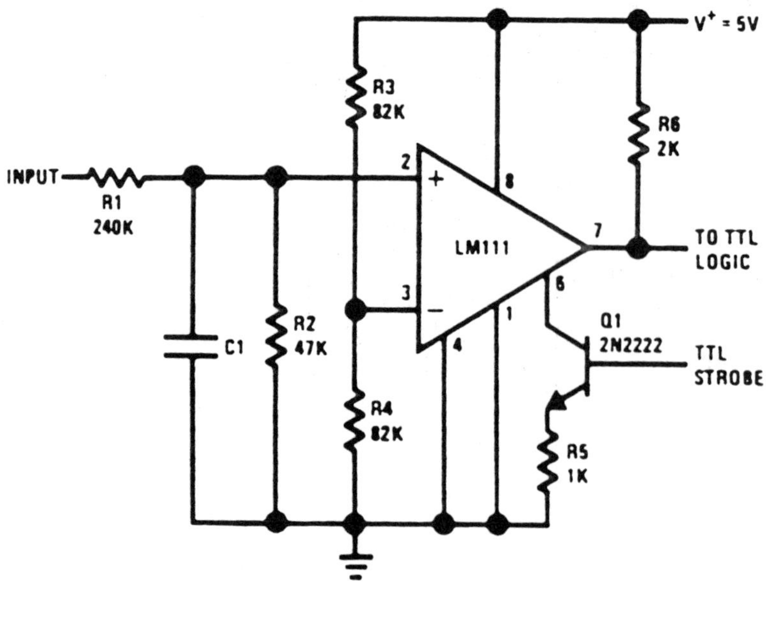

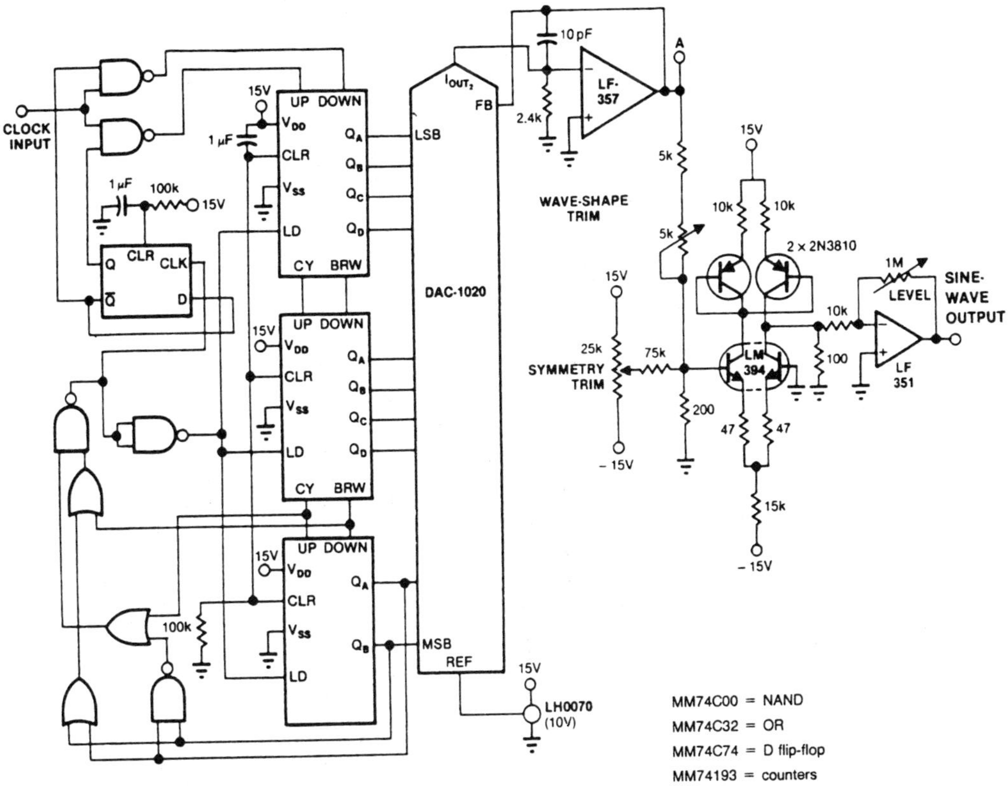

Sine Wave Generator Using Digital Techniques. This unusual circuit uses digital methods to approximate sine wave operation. Such circuits offer the greatest flexibility; however, they are inordinately complex. Here, a 10-bit IC digital-to-analog converter is driven from up/down counters to produce an amplitude-stable triangle current which is fed to the LF357 FET amplifier. This IC, in turn, drives a shaper circuit. The output amplitude of the sine wave is stable and the frequency is solely dependent on the clock used to drive the counters.—"Linear Applications Databook." Reprinted with permission of National Semiconductor.

◀ **High-Level Logic to DTL or TTL Interface.** Using the LM111 voltage comparator IC, this circuit is used for transmitting data between high-level logic and DTL and TTL. The input signal, with 0-V and 30-V logic states, is attenuated to 0 V and 5 V by R1 and R2. Resistors R3 and R4 set up a 2.5-V threshold level for the comparator. This ensures that it switches when the input goes through 15 V. The response time of this circuit can be controlled with C1, if desired, to make it insensitive to fast noise spikes.—"Linear Applications Databook." Reprinted with permission of National Semiconductor.

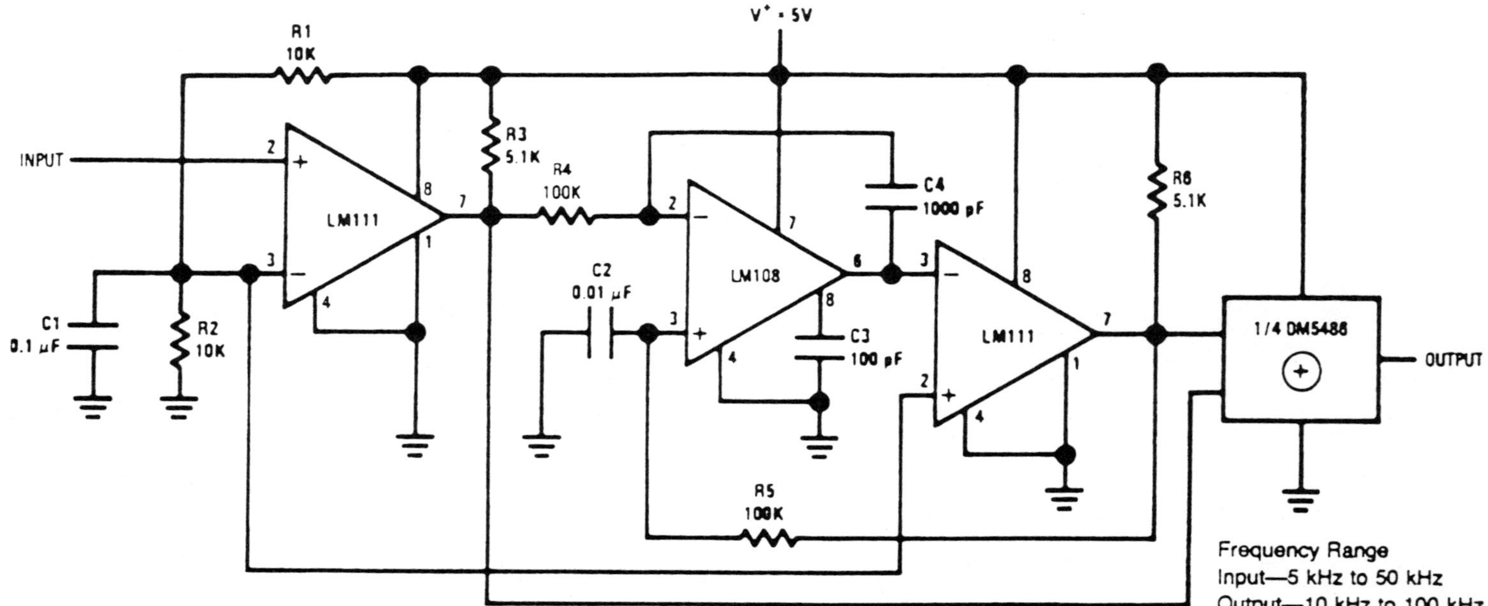

Frequency Range
Input—5 kHz to 50 kHz
Output—10 kHz to 100 kHz

Frequency Doubler. It is a very simple matter to divide by any integer in a digital system. Multiplying by an integer, however, is another case altogether, especially if operation over a wide frequency range is required. This circuit is a frequency doubler that satisfies these requirements. The LM111 comparator is used to shape the input signal and then feed it to an integrator. An LM108 op amp, operating from a 5-V logic supply, serves as this integrator. The output from this IC is a triangle wave, which is fed to a second LM111 comparator. This latter device detects when the waveform goes through a voltage equal to its average value. With the component values shown, this circuit operates at frequencies of from 5 to 50 kHz. If lower-frequency operation is needed, simply increase the values of C1 and C2.—"Linear Applications Databook." Reprinted with permission of National Semiconductor.

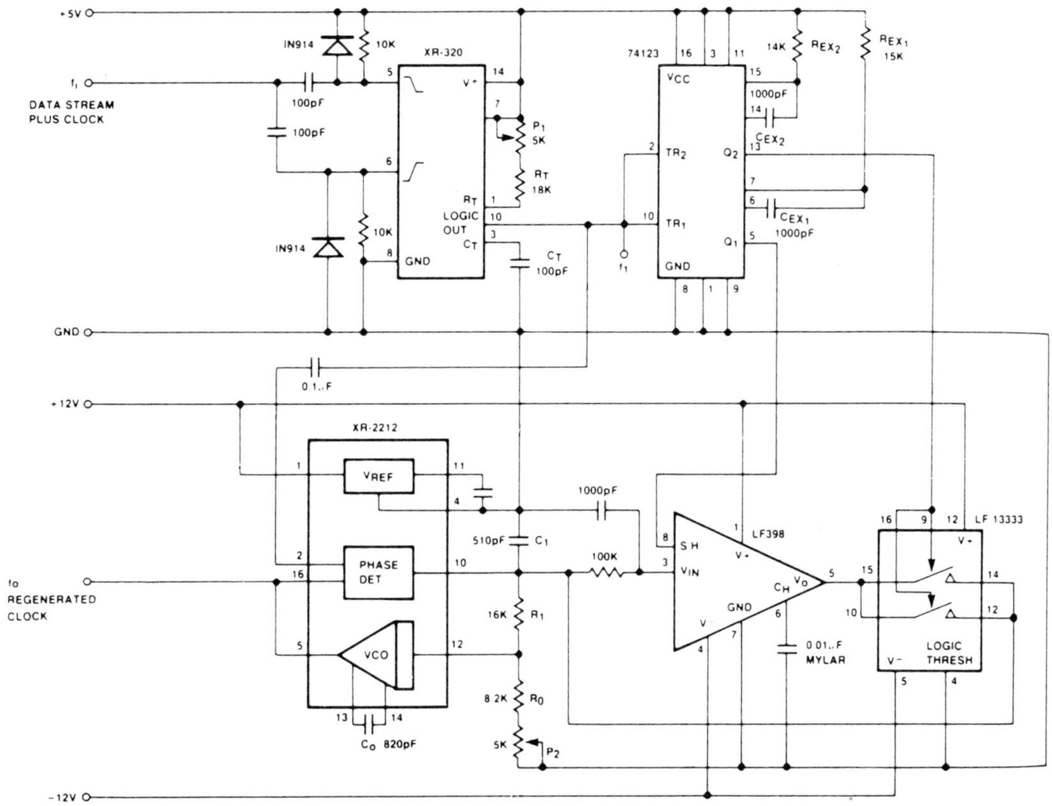

P1 = > ADJUST SO POSITIVE PORTION OF f_i IS EQUAL TO ½ OF THE CLOCK PERIOD
P2 = > ADJUST FOR 90 PHASE SHIFT BETWEEN f1 and fo WITH f_i = f_{CLK}

Clock Recovery System. Recovering encoded serial data from floppy disk systems poses a major design problem as the synchronized clock used to encode the data is embedded within the data stream. This circuit comprises a phase-locked loop system which can be used to recover the clock from the data stream. It uses NRZI protocol with quite excellent stability. This circuit is designed for a clock of 122 kHz. The input system will accept low levels from 0 to 0.5 V and high levels from 1.5 to 5 V. The output provides a 10-V p–p square wave. Calibration is accomplished by adjusting P1 for the output of the XR-320 IC to equal exactly one-half of the clock period of P2 for a 90° phase shift between f_i and f_0 with a constant string of zeros applied at f_i. Incidentally, this same circuit can be used to regenerate or clean up a clock with occasional missing cycles by applying it to the point labeled (f_i) and eliminating the XR-320 from the circuit.—"EXAR Databook." Reproduced with the express authorization of EXAR Corporation.

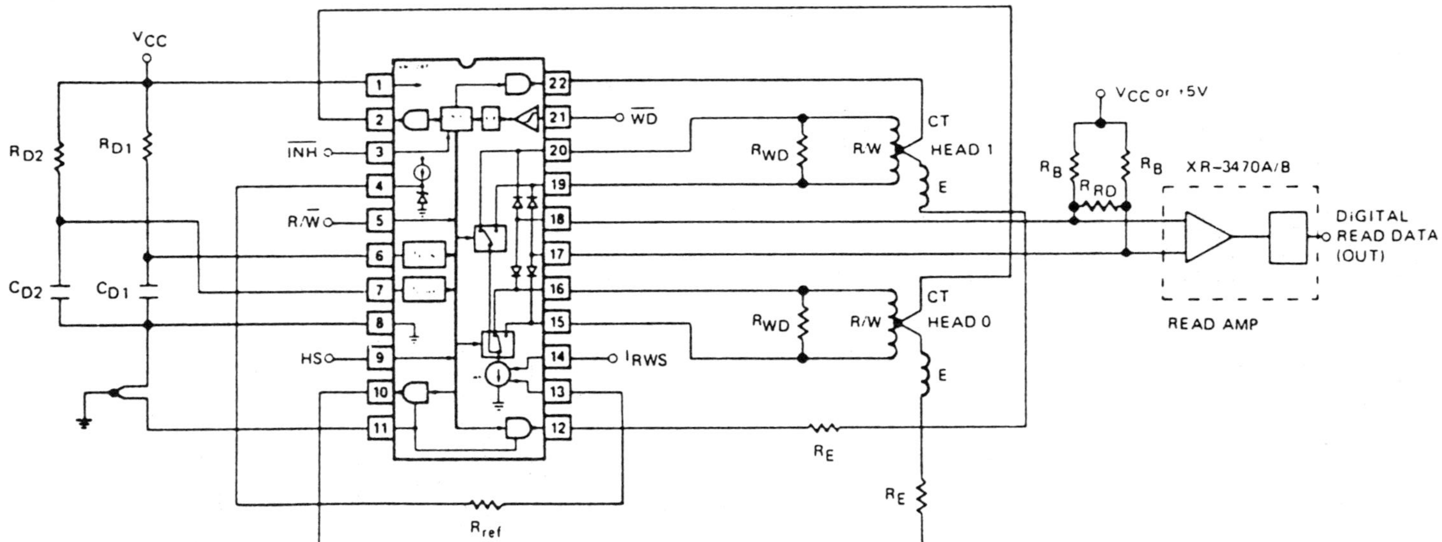

Dual Head Floppy Disk Write System. This circuit uses the EXAR XR-2247 floppy disk amplifier IC. The external components are calculated as follows. Write current

$$I_{wr} = (5.3) * ((V_{ref} - V_{mir}) / R_{ref}) * I_{rws} = \text{low}$$

Given $I_{wr} = 4.1$ mA and $R_{ref} = 10$ kΩ, then I_{ref} (the current into pin 13) should not exceed 2.0 mA. Erase current

$$I_e = (V_{cth} - V_{oel}) / RE = (V_{cc} - 2\text{ V}) / RE$$

Given $I_e = 50$ mA and $V_{cc} = 12$ V, then RE = 200 Ω.

Control of the erase outputs can also be handled from an external source by grounding pin 6 and driving pin 7 directly. Resistors labeled R_{WD} are used to damp any ringing that may occur when the write current transitions are applied to the head. Their value is determined by the head characteristics and the amount of damping desired. R_{RD} provides additional damping in the read mode, if this is desired.—"EXAR Databook." Reproduced with the express authorization of EXAR Corporation.

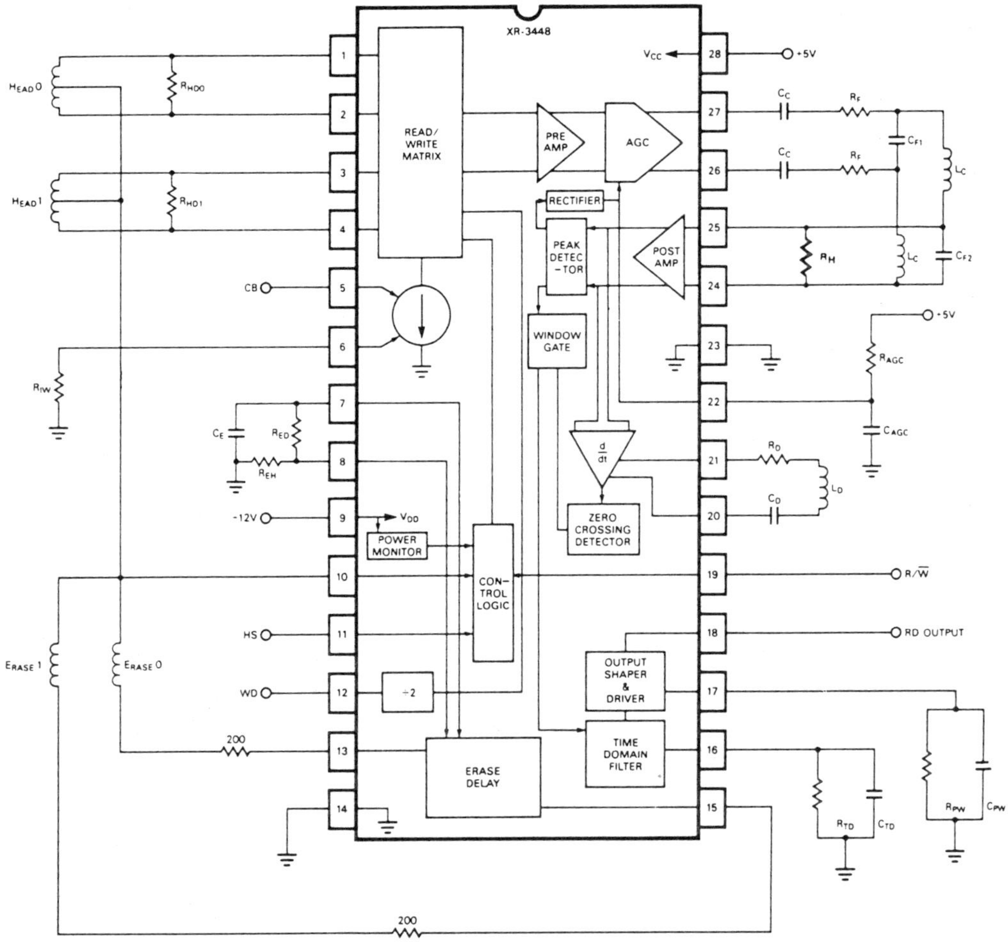

TYPICAL COMPONENT VALUES

Component	Typical Value	Recommended Range	Component	Typical Value
R_{IW}	560Ω	180 - 680Ω	R_H	1.5kΩ
R_{E0}, R_{E1}	280Ω	100 - 680Ω	C_C	.022μF
R_{ED}, R_{EH}	10kΩ	5k - 30kΩ	R_F	470Ω
C_E	.047μF	0.01 - 0.068μF	C_{F1}	.001μF
R_{PW}	10k	5k - 25kΩ	C_{F2}	470pF
C_{PW}	100 pF	51 pF - 1000 pF	L_C	680μH
C_{TD}	100pF	51 pF - 330 pF	R_D	200Ω
R_{AGC}	6.81k	3.3k - 25kΩ	C_D	1000pF
C_{AGC}	1μF	.01μF - 1μF	L_D	56μH

Floppy Disk Read System. Here is a basic floppy disk read circuit to accompany the previous write circuit. If maximum read gain is needed, try deleting R_{AGC} and C_{AGC} and ground pin 22.—"EXAR Databook." Reproduced with the express authorization of EXAR Corporation.

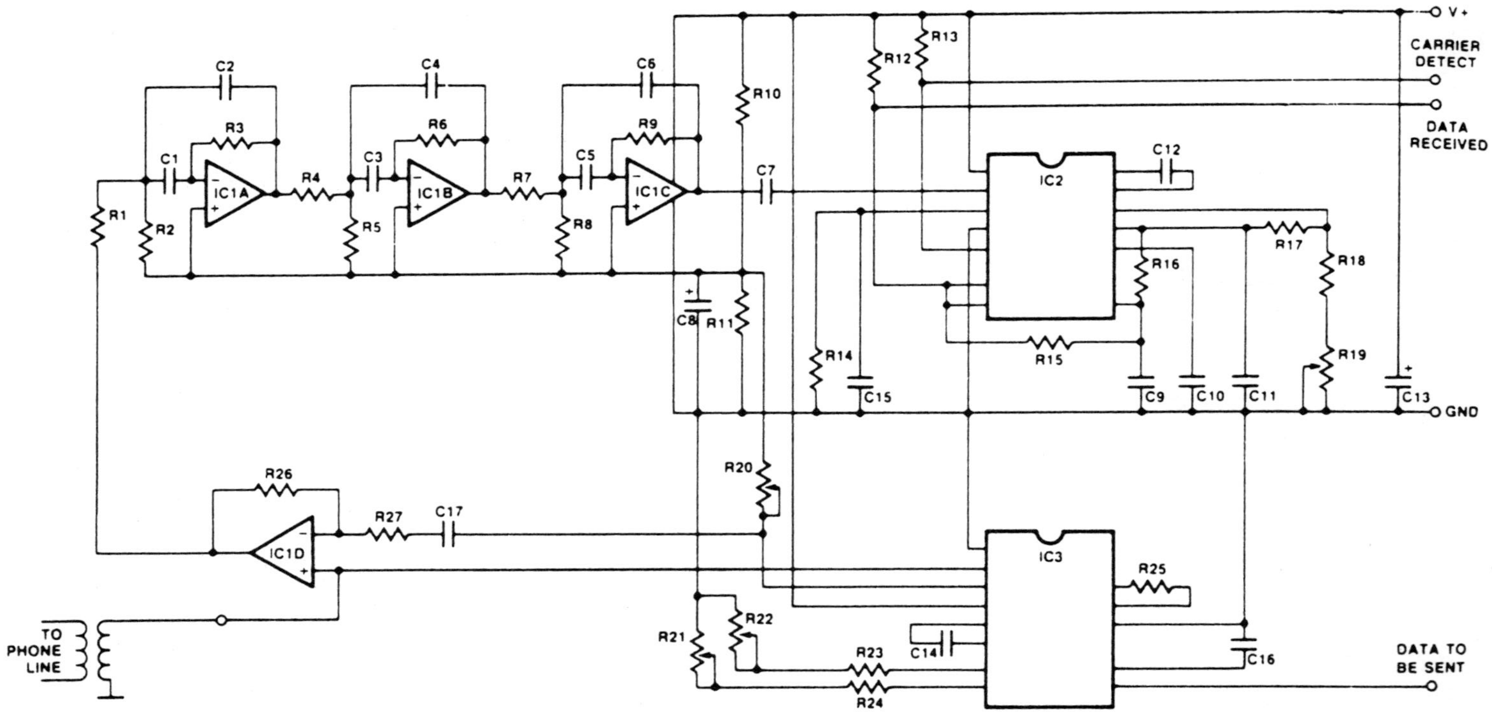

Complete FSK Modem. Using the EXAR XR-2211 as a demodulator and the XR-2206 as a modulator, a complete frequency shift keying modem can be easily constructed with a bare minimum of discrete components. This circuit has been designed for 12-V DC operation. The data inputs accept TTL compatible signal levels, while the outputs provide 0- to +12-V signal levels. To calibrate this circuit, first adjust the modulator section with a low signal on its input. R21 is adjusted for 1270 Hz or 2225 Hz for originate and answer, respectively. Then, with a high signal on the input, adjust R22 for 1070 Hz or 2025 Hz. The demodulator is aligned by feeding an alternating 1070 and 1270 Hz or 2025 and 2225 Hz signal to the modem input in a square wave fashion. The modulating frequency should be 150 Hz, which is one-half the system baud rate of 300. R20 is used to set the modulator output level.—"EXAR Databook." Reproduced with the express authorization of EXAR Corporation.

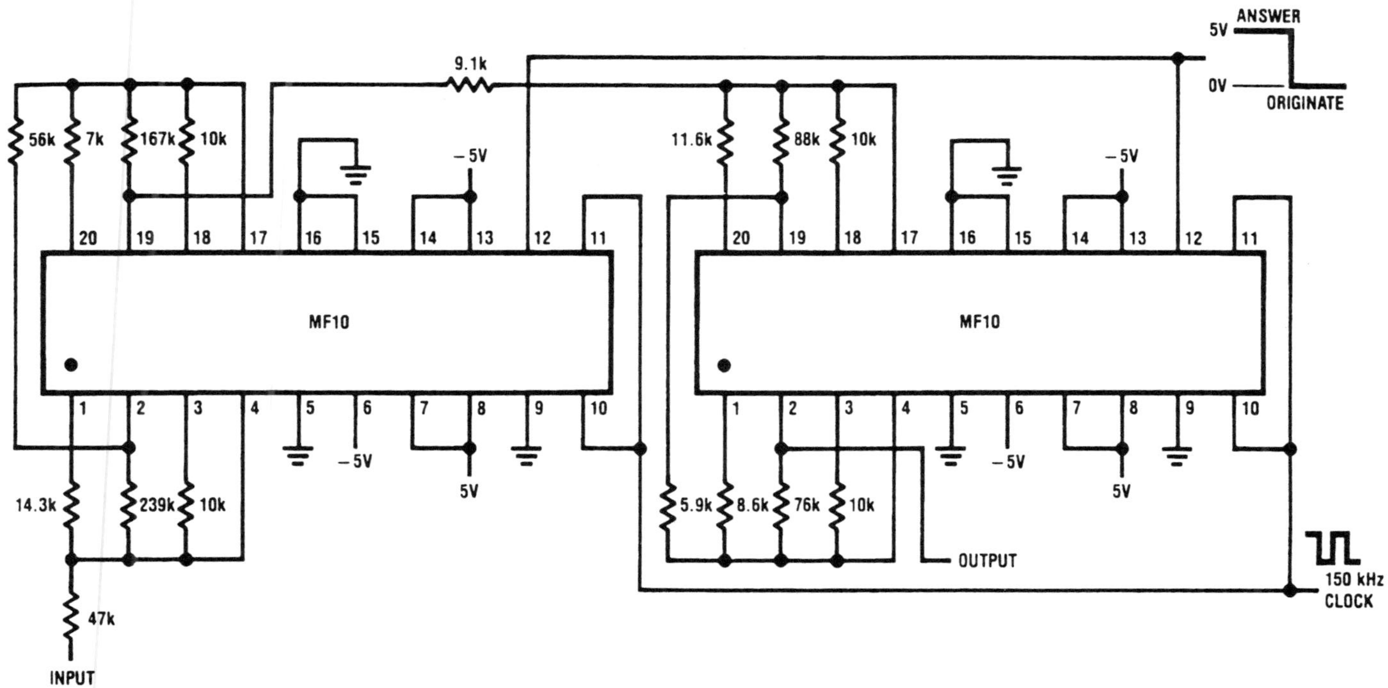

Full-Duplex 300 Baud Modem Filter. This circuit is an excellent example of the simplicity and ease with which highly efficient filters may be constructed from IC filter building blocks such as the MF10 shown here. This filter is an 8th-order, 1-dB ripple Chebyshev bandpass which functions as both an 1170-Hz originate filter and a 2125-Hz answer filter. Control of each operation is set by the logic level at pins 12. This is the 50/100/CL input, so only one clock frequency is required. The overall filter gain is 22 dB. This circuit lends itself to highly compact construction on a small piece of printed circuit board.—"Linear Applications Databook." Reprinted with permission of National Semiconductor.

Chapter 6

Circuits for the Workbench

This chapter includes a handful of circuits that were, quite frankly, left over. They do have a common thread and for this reason are not lumped with the miscellaneous circuits near the end of this book. The circuits featured here have very definite applications on the technician's test bench or around the bench. Included here are a touch switch, several converters and generators, and a few other obscure gadgets that might have some esoteric value.

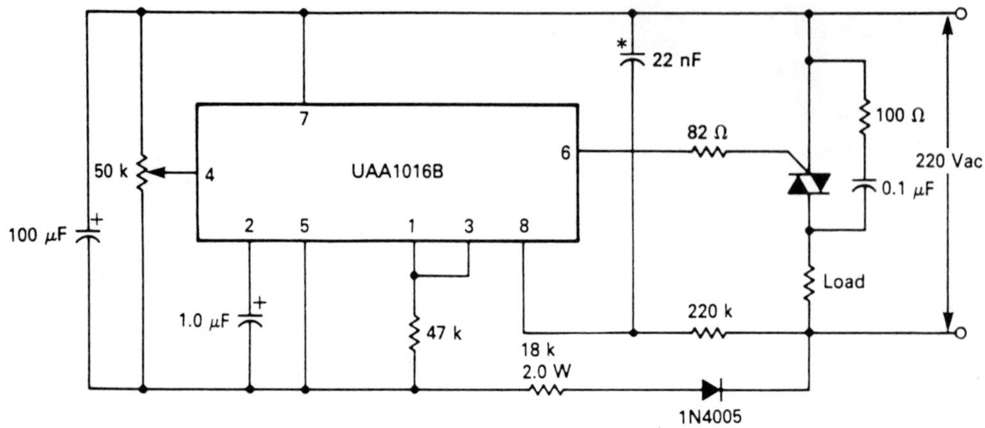

Electronic Rheostat. The Motorola UAA1016B is a zero-voltage controller designed specifically to drive TRIACs with the zero-voltage technique that allows RFI-free power regulation of resistive loads. In this simple application, the chip is used as an electronic rheostat, having 100% efficiency (due to the zero-voltage technique). Note that the 22-nF capacitor that is paralleled across the output is necessary only if the load has an inductive component.—"Linear and Interface Integrated Circuits." Copyright of Motorola, Inc. Used by permission.

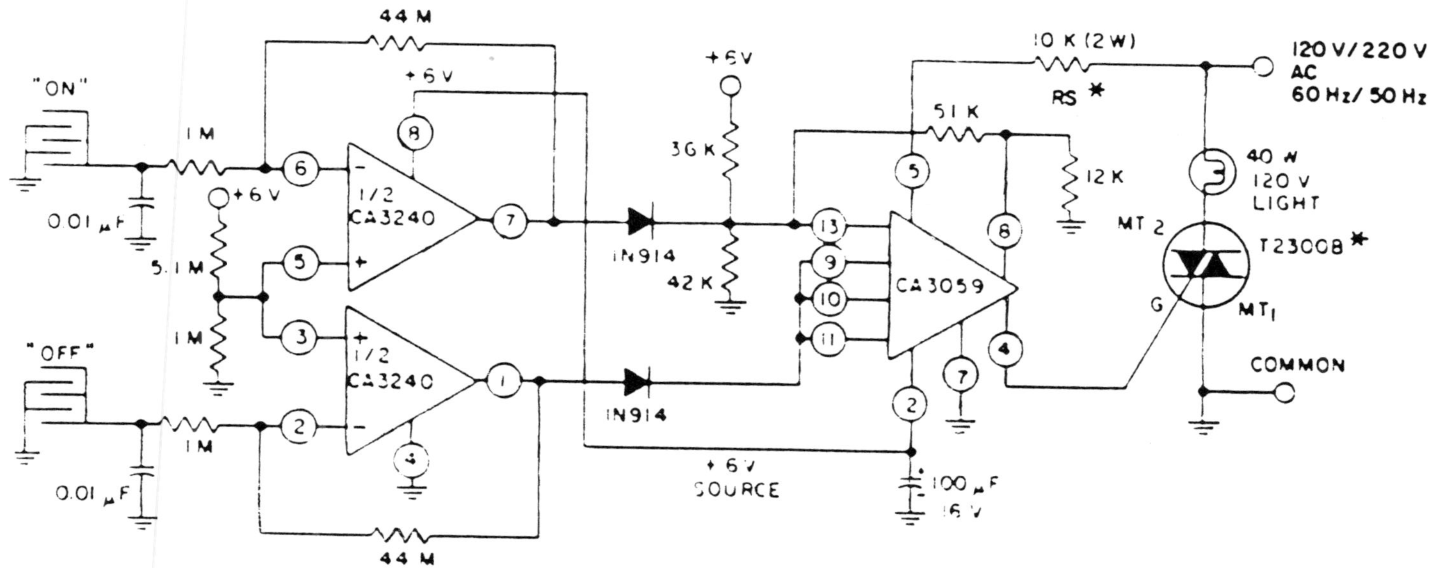

*AT 220 V OPERATION, TRIAC SHOULD BE T23000, RS = 18 K, 5 W

On–Off Touch Switch. This simple circuit has possible applications as a part of intrusion alarm circuits or simply as a silent means of switching other electrical circuits on and off. It uses a dual op amp IC to send small currents flowing between two contact points on touch plates made from PC boards. When the *on* plate is touched, current flows between the two halves of the grid. This causes a positive shift in the output voltage of the top amplifier. These positive transitions are fed to the input of the CA3059, which is used as a latching circuit and zero-crossing TRIAC driver. When a positive pulse occurs at pin 7 of the op amp, the TRIAC is turned on and held on by the CA3059 driver circuit. When a positive pulse occurs at terminal 1 of the dual op amp, the TRIAC is turned off and held off in a similar manner. The dual op amp derives its power from the internal power supply of the CA3059. This device (CA3059) is a zero-voltage switch designed for thyristor control applications.—"RCA Solid State Databook—Integrated Circuits for Linear Applications." Courtesy of Harris Semiconductor.

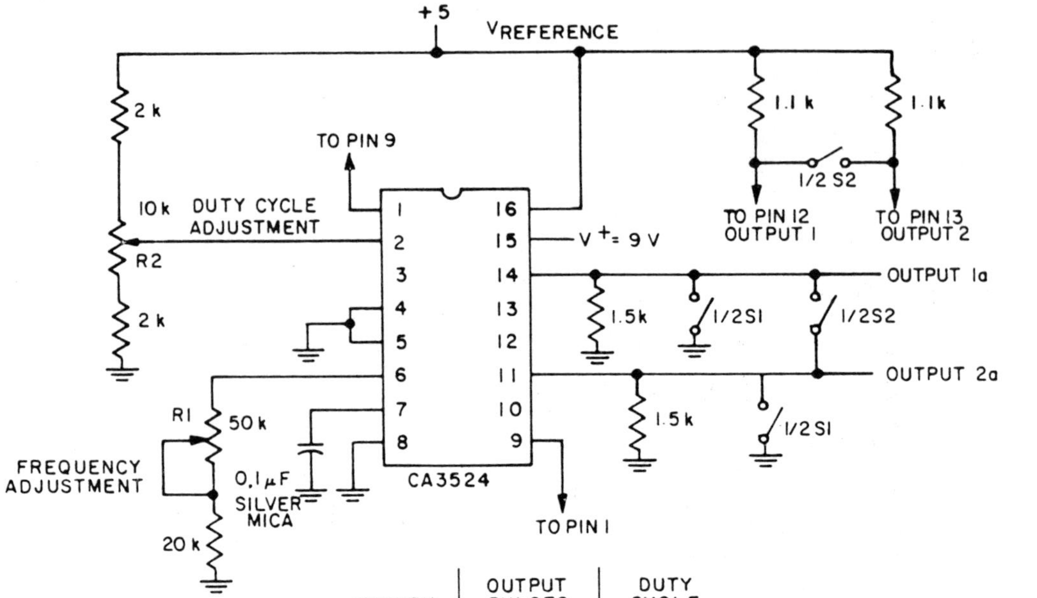

SWITCH	OUTPUT PULSES	DUTY CYCLE
S1	0 -5 V	0 – 45 %
S2	–	0 – 90 %

Low-Frequency Pulse Generator. This circuit uses a regulating pulse-width modulator IC as the heart of a low-frequency pulse generator. All of the major components for such a circuit are already contained on the IC. These include error amplifier oscillator, oscillator reference regulator, and output transistor drivers. A regulated 5-V pulse of 0 to 45% *on* time is possible over a frequency range of 150 to 500 Hz. S1 is used to go from a 5-V output pulse (S1 closed) to a 2.5-V output pulse (S1 open) with a duty cycle range of 0 to 45%. The output frequency will be approximately half of the oscillator frequency when the output transistors are not connected in parallel. S2 will allow both output stages to be parallel for an effective duty cycle of 0 to 90%, with the output frequency range from 150 to 500 Hz. In this configuration, the 50-kΩ potentiometer (R1) controls the frequency, while a 10-kΩ potentiometer (R2) determines the duty cycle.—"RCA Solid State Databook—Integrated Circuits for Linear Applications." Courtesy of Harris Semiconductor.

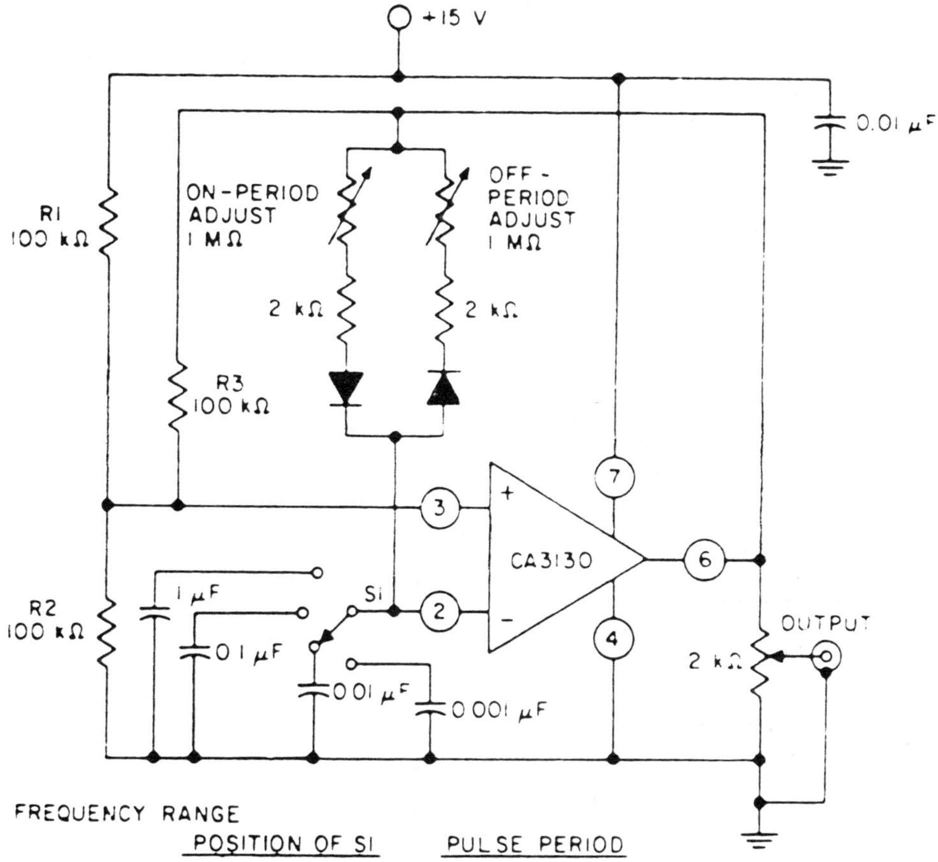

FREQUENCY RANGE

POSITION OF S1	PULSE PERIOD
0.001 μF	4 μs TO 1 ms
0.01 μF	40 μs TO 10 ms
0.1 μF	0.4 ms TO 100 ms
1 μF	4 ms TO 1 s

Pulse Generator with Independent Duty Cycle Control. The CA3130 op amp from RCA offers an exceptionally high resistance, making it very suitable for multivibrator circuit design. This is due to the fact that timing circuits with high R/C ratios can be used in this application. This pulse generator is an astable multivibrator with provisions for independent control of the *on* and *off* periods. R1 and R2 are used to bias the op amp to the midpoint of the supply voltage, and R3 is the feedback resistor. The pulse repetition rate is selected by switching S1 to the desired position, and the rate remains essentially constant when the resistors that determine *on/off* period are adjusted.—"RCA Solid State Databook—Integrated Circuits for Linear Applications." Courtesy of Harris Semiconductor.

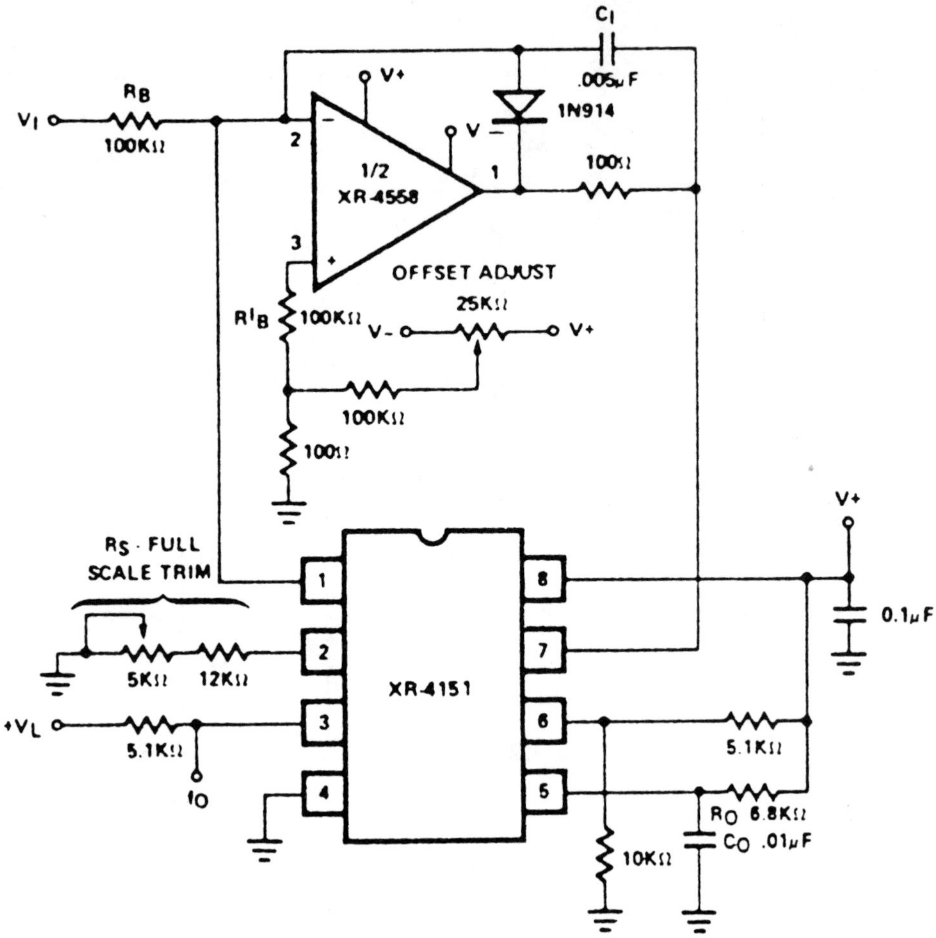

Precision Voltage-to-Frequency Converter. Using the XR-4151 voltage-to-frequency converter IC, this application also incorporates an op amp integrator to provide typical linearity characteristics of 0.05% over the range of 0 to −10 V. Offset is adjustable to zero, and, unlike many VFC designs that lose linearity below 10 mV, this circuit retains its linearity over the full range of input voltage values, all the way to 0 V. Trim the full scale adjust potentiometer at $V_i = -10$ V. The offset pot should be set for 10 Hz with an input voltage of −10 mV.—"EXAR Databook." Reproduced with the express authorization of EXAR Corporation.

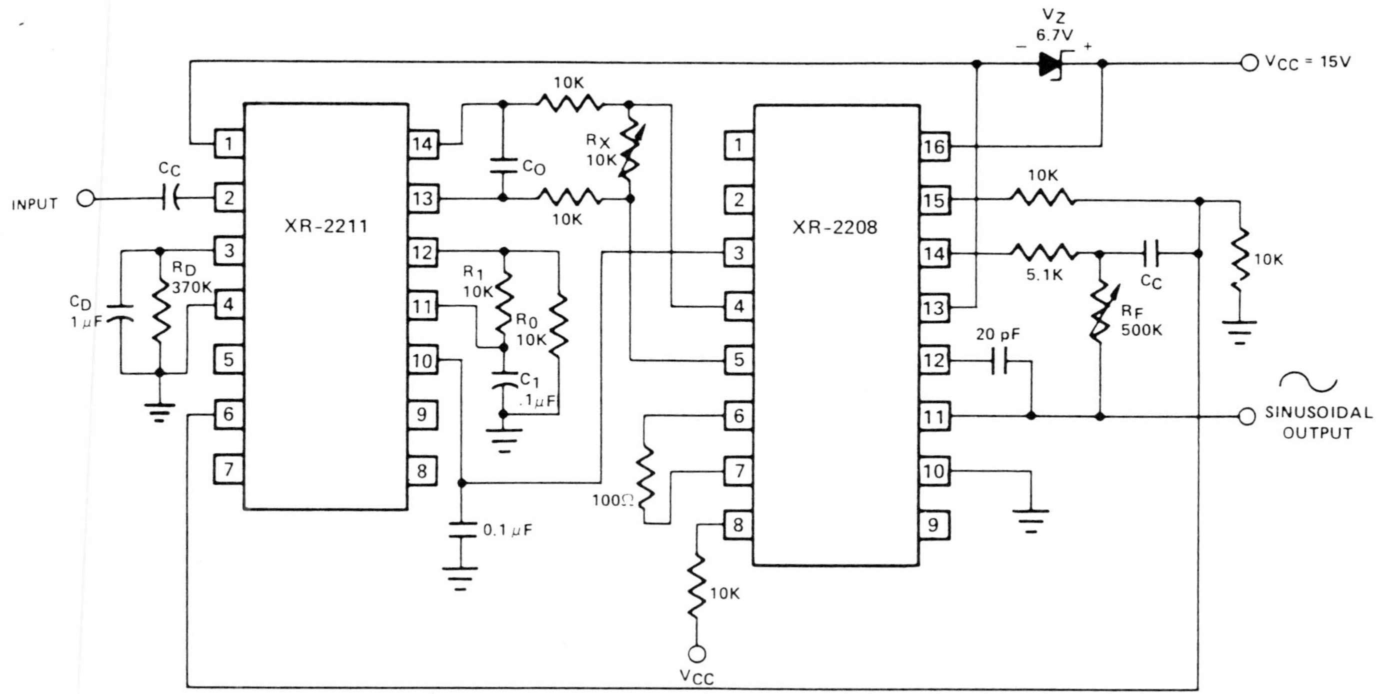

R_X = Distortion Adj. Potentiometer
R_F = Output Amplitude Adj. Pot.

C_C = Coupling Capacitor ($\geqslant 0.1\ \mu F$)

Universal Sine Wave Converter. Utilizing the EXAR XR-2211 and XR-2208 ICs, this circuit operates from a 15-V DC power supply and converts square wave inputs into pure sine wave signals. The distortion of the sine wave is minimized by adjusting R_x, which sets the output amplitude of the XR-2211's triangle wave oscillator. The minimum input level required for circuit operation is 10 mV (rms). This circuit will generate a (nearly) sinusoidal output with input signals from very low frequencies up to approximately 100 kHz.—"EXAR Databook." Reproduced with the express authorization of EXAR Corporation.

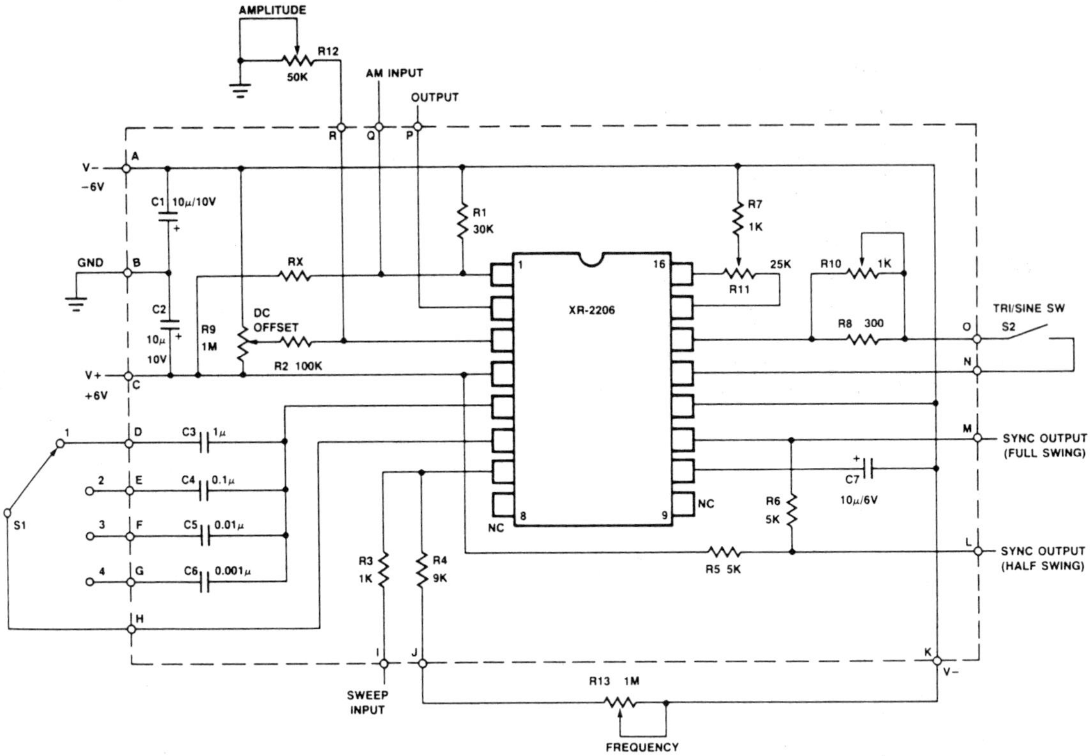

NOTE:

1. For Single Supply Operation Lift GND Connection Keeping R12 Across Terminals R and B Intact, and Connect Terminal A to GND.

2. For Maximum Output, R_X may be open. R_X = 68 KΩ is Recommended for External Amplitude Modulation.

High-Quality Function Generator. Using the XR-2206 function generator IC, this circuit is designed to operate with either a single-polarity 12-V DC power supply or a 6-V DC bipolar supply. For most applications bipolar supply operation is preferred, since it results in an output DC level that is nearly at ground potential. This circuit provides three basic waveforms, sine, triangle, and square wave. There are four overlapping frequency ranges which provide coverage of from 1 Hz to 100 kHz. In each range, the frequency may be varied over a 100 : 1 tuning ratio. The sine or triangle output can be varied from zero to over 6 V p–p from a 600-Ω source at the output terminal. High-quality variable controls are essential to the smooth operation of this circuit.—"EXAR Databook." Reproduced with the express authorization of EXAR Corporation.

Chapter 7

RF Reception Circuits

The circuits featured in this section address the radio-frequency spectrum in some way and are mostly limited to broadcast applications (i.e., AM and FM receivers, detectors, demodulators, and so on). This area of electronics design has been greatly aided by the development of highly specialized linear integrated circuits that just about do it all. Today, it is possible to purchase what amounts to a 95% complete AM or FM receiver on a single chip. This would seem to be the most perfect of all situations in that cost is generally held to a bare minimum, circuit dependability is as high as possible due to the small number of discrete components and interconnections, and design/construction criteria are minimal.

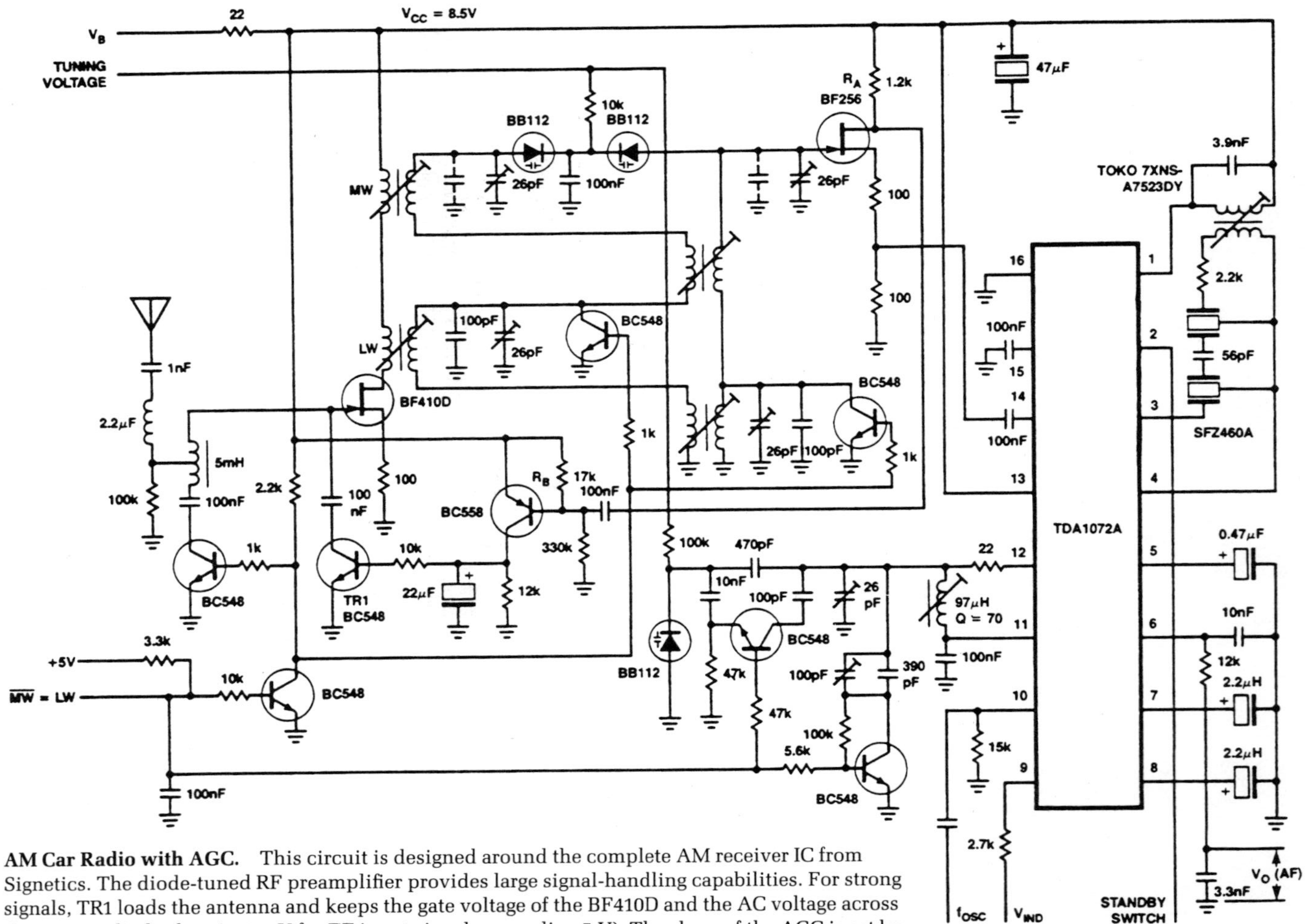

AM Car Radio with AGC. This circuit is designed around the complete AM receiver IC from Signetics. The diode-tuned RF preamplifier provides large signal-handling capabilities. For strong signals, TR1 loads the antenna and keeps the gate voltage of the BF410D and the AC voltage across the vericap diodes low (130 mV for RF input signals exceeding 5 V). The slope of the AGC is set by R_a and the onset of gain control by R_b. Because the RF gain control is derived from the output of the tuned RF preamplifier, there is no masking of desired weak signals situated close to strong ones.—"Linear Data Manual Volume 1: Communications." Courtesy of Signetics Company, a division of North American Philips Corporation.

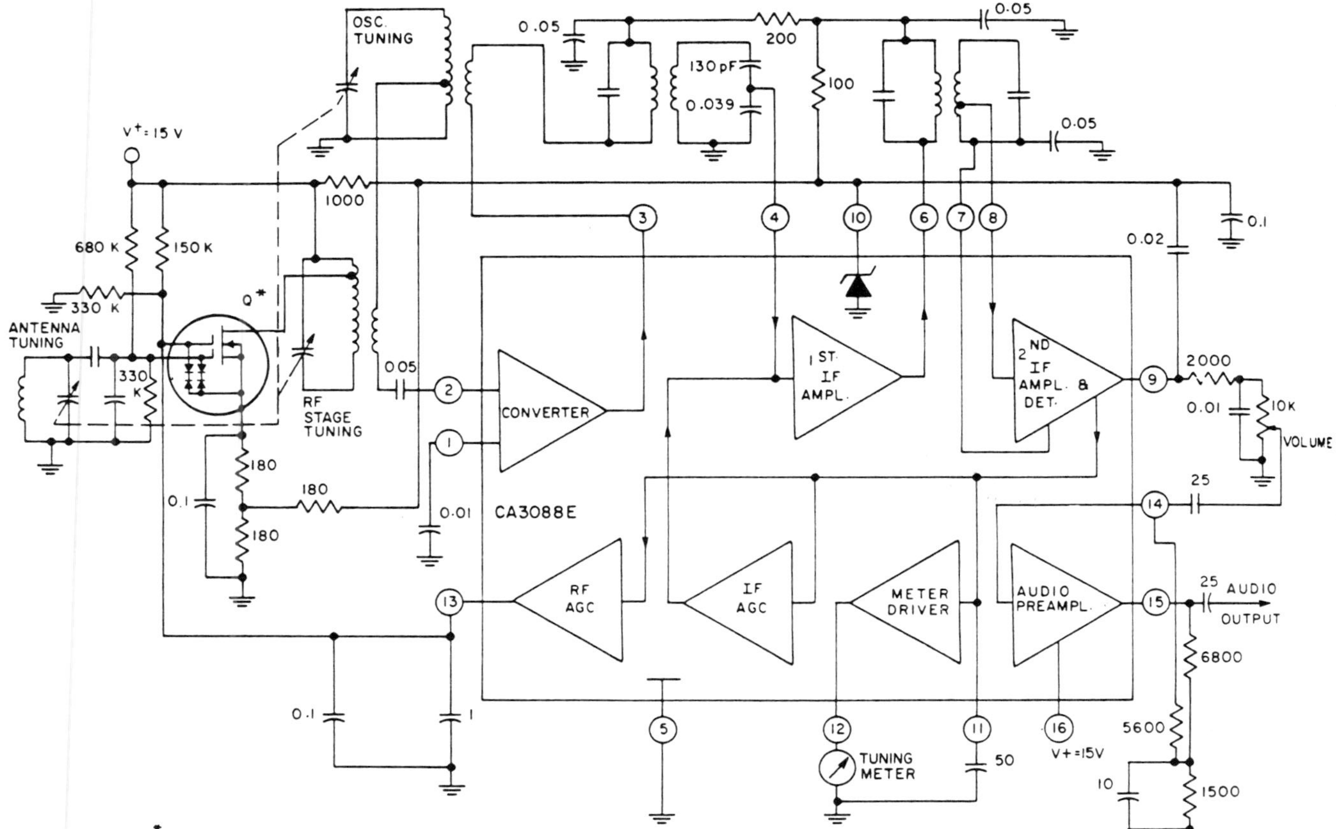

*Q = RCA-40841 (DUAL GATE-PROTECTED MOS/FET)

AM Broadcast Receiver. This will probably be relegated into the "worthless circuit" category, but it is included here for those readers who have some need or desire to build an AM radio receiver with an optional RF amplifier stage. The heart of the circuit is a CA3088 AM receiver subsystem from RCA. On-chip features include an AM converter, IF amplifiers, detector, and audio preamplifier. If desired, the preamplifier portion can be dropped altogether. All in all, though, it would probably be far easier to simply purchase a completed AM radio module, of which many abound in the electronics marketplace.—"RCA Solid State Databook—Integrated Circuits for Linear Applications." Courtesy of Harris Semiconductor.

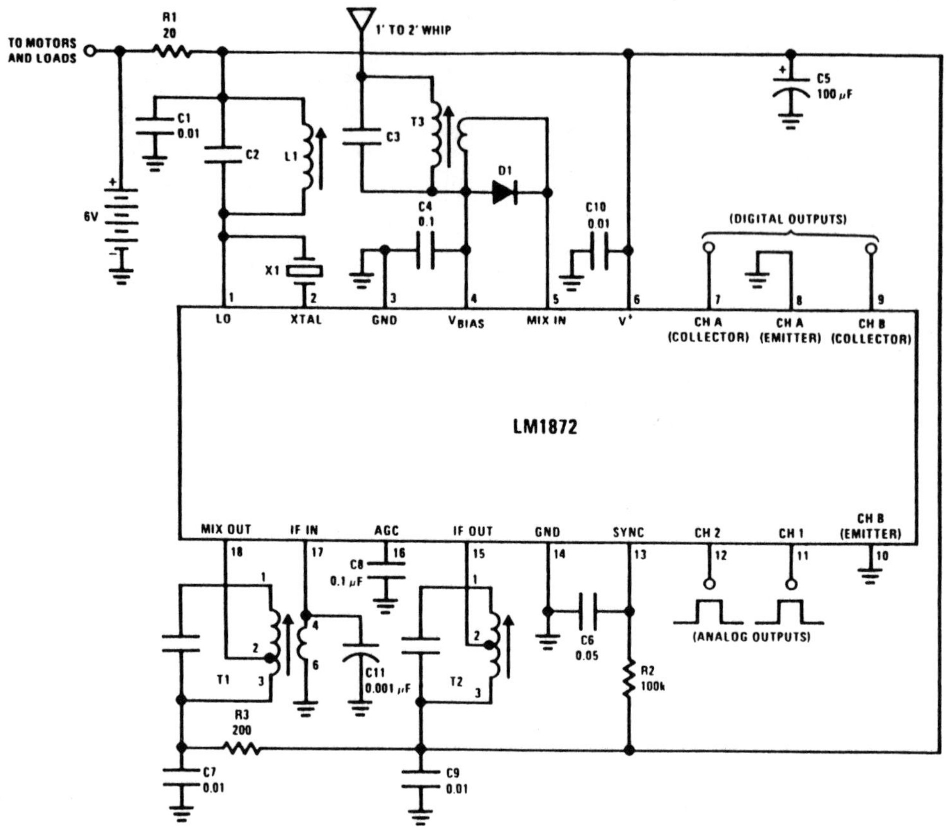

R1 — Motor decoupling

R2 — Sync timer; $R2 = \frac{t'_{SYNC}}{0.7\,C6}$, $R2 \le 470k$

R3 — Mixer decoupling

C1 — LO bypass; optional

C2 — LO tank; C2 = 43 pF @ 27 MHz
= 24 pF @ 49 MHz

C3 — Ant. input tank; C3 = 39 pF @ 27 MHz
= 24 pF @ 49 MHz

C4 — V_{BIAS} bypass

C5 — Motor decoupling

C6 — Sync timer; $C6 = \frac{t'_{SYNC}}{0.7\,R2}$, $C6 \le 0.5\ \mu F$

C7 — Mixer decouple; $0.01\ \mu F \le C7 \le 0.1\ \mu F$

C8 — AGC

C9 — IF bypass; optional

C10 — V^+ bypass; $0.01\ \mu F \le C10 \le 0.1\ \mu F$

C11 — LO bypass

L1 — LO coil
Toko* 10k type (KXNA-4434 DZ) 9T; 0.8 μH @ 27 MHz
Toko* 10k type (KEN-4028 DZ) 6T; 0.4 μH @ 49 MHz
L1 could be made a fixed coil, if desired.

T1 — 455 kHz mixer transformer
Toko* 10 EZC type (RMC-202313 NO), Qu = 110
Pin 1–2, 131T; pin 2–3, 33T
Pin 1–3, 164T; pin 4–6, 5T

T2 — 455 kHz IF transformer
Toko* 10 EZC type (RMC-402503 NO), Qu = 110
Pin 1–2, 98T; pin 2–3, 66T
Pin 1–3, 164T; pin 4–6, 8T

T3 — Ant. input transformer
Toko* 10k type (KXNA-4434 DZ), 3T sec. & 9T pri. of 0.8 μH @ 27 MHz
Toko* 10k type (KEN-4028 DZ), 1½T sec. & 6T pri. of 0.4 μH @ 49 MHz

X1 — 3rd overtone parallel-mode crystal

D1 — Electrostatic discharge (ESD) protection

*Toko America
1250 Feehanville Drive
Mount Prospect, IL 60056
(312) 297-0070

◀ **Radio Control Receiver.** This circuit is intended for use with encoding circuits designed for remote radio control of various loads. The receiver circuit configuration is a simple single-conversion design with AGC, which mixes down to 455 kHz and provides 58 dB of gain using the components shown here. The active digital detector provides an additional 30-dB gain over a silicon diode. This results in an overall system gain of approximately 88 dB. If more or less gain is desired, then different transformers can be utilized. The frequency range of operation is extremely broad and runs from 50 kHz to 72 MHz. This is a description of the chip range. The actual circuit depicted will operate at 27 MHz or at 49 MHz. Due to the large number of external components, a separate component chart is provided.—"Linear Databook III." Reprinted with permission of National Semiconductor.

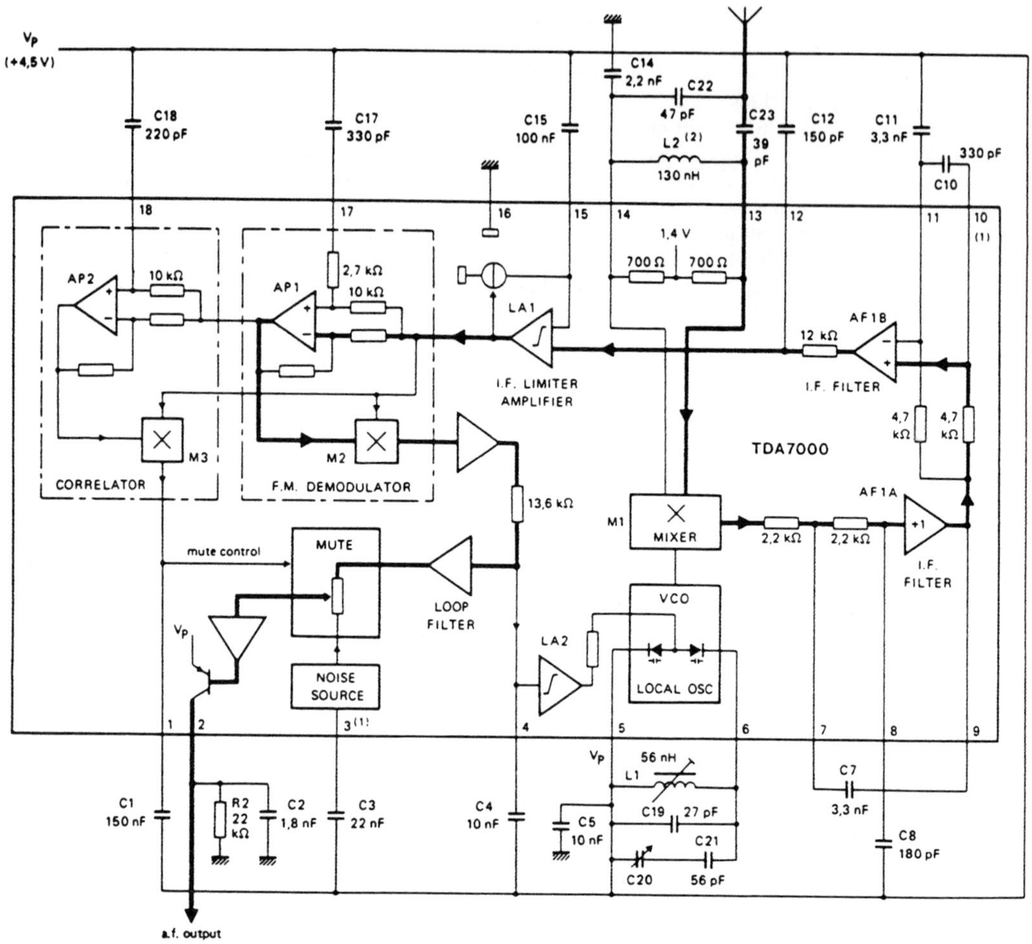
Vp
(+4,5 V)
C18
220 pF
C17
330 pF
C15
100 nF
C14
2,2 nF
C22
47 pF
L2 (2)
130 nH
C23
39 pF
C12
150 pF
C11
3,3 nF
330 pF
C10
AP2
10 kΩ
AP1
2,7 kΩ
10 kΩ
LA1
1,4 V
700 Ω
700 Ω
12 kΩ
AF1B
I.F. LIMITER AMPLIFIER
I.F. FILTER
4,7 kΩ
4,7 kΩ
M3
M2
TDA7000
CORRELATOR
F.M. DEMODULATOR
AF1A
13,6 kΩ
M1
MIXER
2,2 kΩ
2,2 kΩ
+1
I.F. FILTER
mute control
MUTE
LOOP FILTER
VCO
LA2
LOCAL OSC
NOISE SOURCE
Vp
56 nH
L1
C19
27 pF
C5
10 nF
C21
C20
56 pF
C7
3,3 nF
C8
180 pF
C1
150 nF
R2
22 kΩ
C2
1,8 nF
C3
22 nF
C4
10 nF
a.f. output

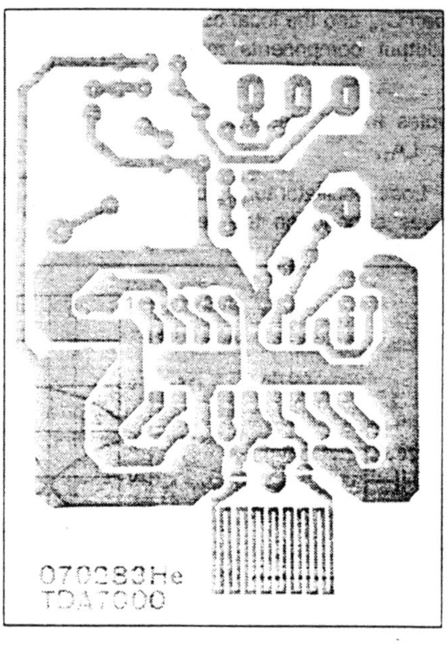
070283He
TDA7000

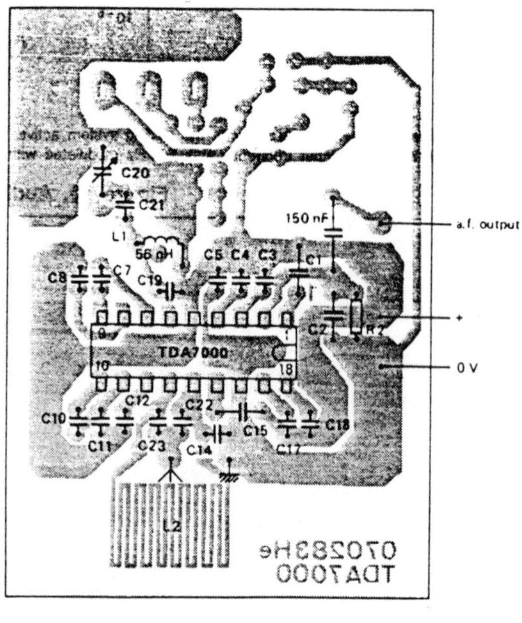
C20
C21
L1
56 nH
150 nF
a.f. output
C5 C4 C3
C1
C8
C7
C19
C2
R2
+
0 V
TDA7000
9
10
18
C12
C22
C10
C11
C23
C14
C15
C17
C18
L2
070283He
TDA7000

◀ **Complete FM Radio on a Chip.** Using the Signetics TDA7000 along with a small handful of capacitors and a few inductors, a complete FM radio can be constructed. This chip consists of a local oscillator, mixer, two-stage active IF filter followed by an IF limiter/amplifier, a quadrature FM demodulator, and an audio muting circuit controlled by an IF waveform correlator. A printed circuit board layout is also provided to facilitate construction.—"Linear Data Manual Volume 1: Communications." Courtesy of Signetics Company, a division of North American Philips Corporation.

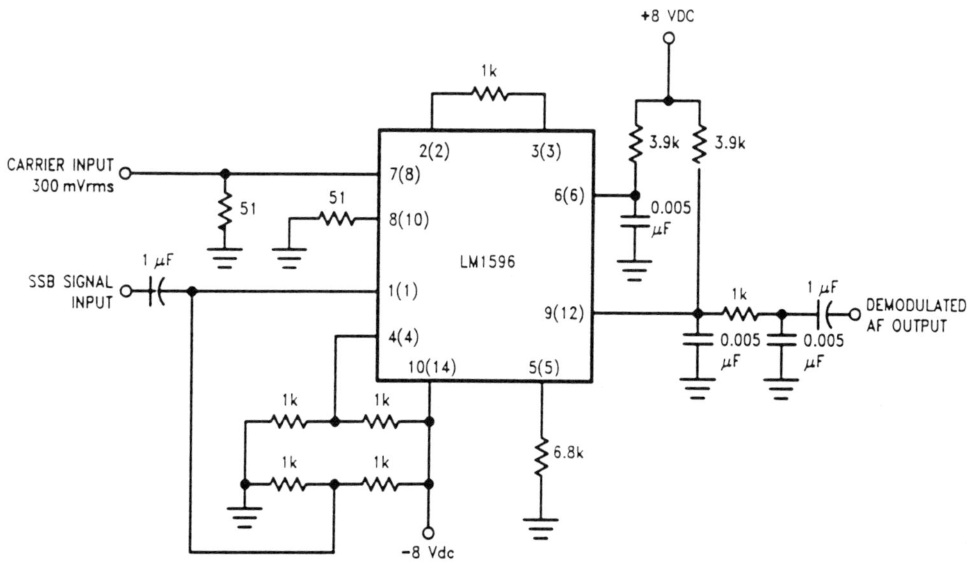

SSB Suppressed Carrier Demodulator. This demodulator or product detector circuit uses the LM1596 balanced modulator/demodulator IC from National Semiconductor. The carrier signal is applied to the carrier input port with sufficient amplitude for switching operation. A carrier level input of 300 mV is about optimum. The composite SSB signal is applied to the signal input port with an amplitude of 5.0 to 500 mV. All output signal components except the desired demodulated audio are filtered out, so that an offset adjustment is not required. Incidentally, this same circuit may also be used as an AM detector simply by applying composite and carrier signals in the same manner as described for the product detector.—"Linear Databook III." Reprinted with permission of National Semiconductor.

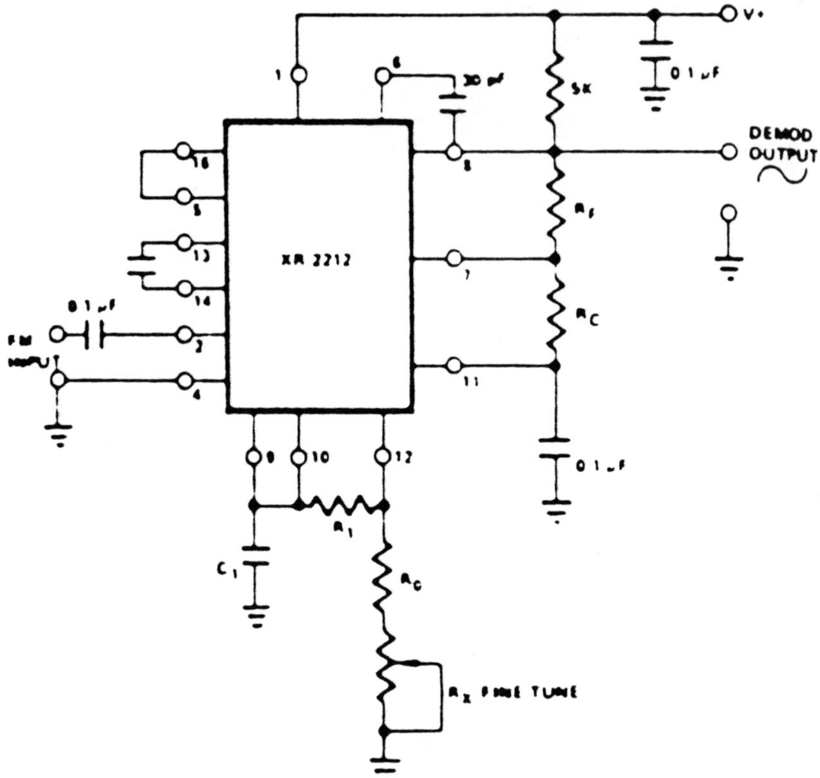

FM Demodulator. The VCO output at pin 5 is directly connected to the phase detector at pin 16. The demodulated signal is obtained at pin 10 which is the output of the phase detector.The op amp section of this IC is used as the buffer amplifier to provide both additional voltage amplification and current drive capability. R_0C_0 values set the VCO center frequency. R1 sets the tracking bandwidth and C1 sets the low-pass filter time constant. The VCO center frequency should be set to that of the FM carrier.—"EXAR Databook." Reproduced with the express authorization of EXAR Corporation.

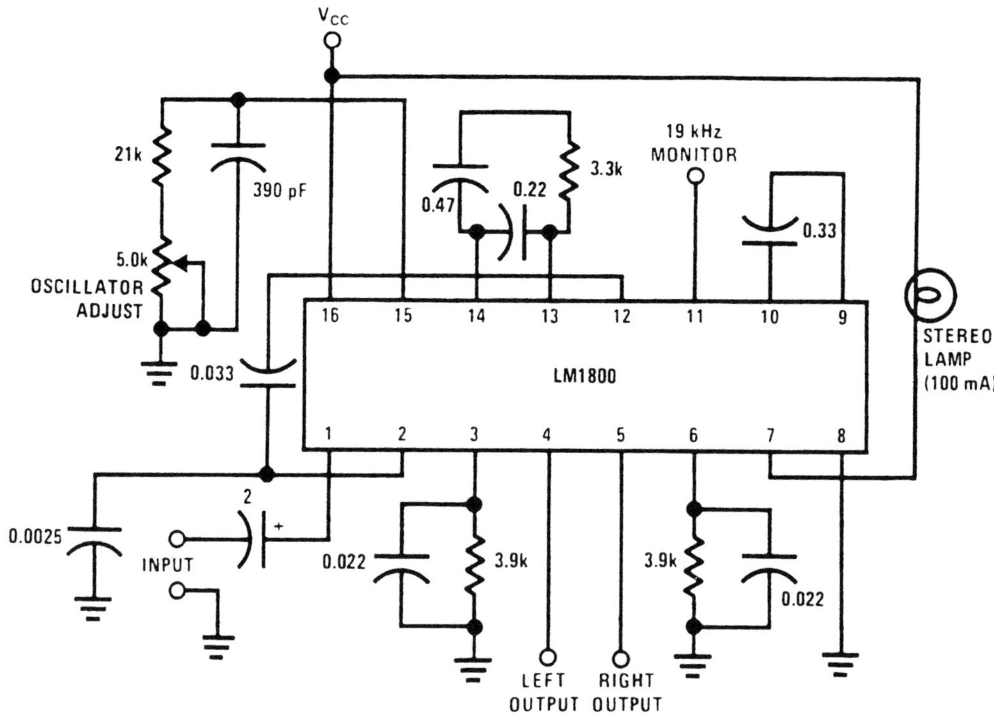

PLL FM Stereo Demodulator. This complete stereo demodulator consists of only 14 discrete components and the NS LM1800 phase-locked loop demodulator IC. The circuit can be driven by a wide range of supply voltages with 12 V DC being about optimal. This is a very basic circuit with absolutely no frills, but it does demonstrate the simplicity with which typically complex circuits can be constructed using today's "packed" ICs. This chip is designed to deliver high-fidelity sound while still coming in at a reasonable cost. It is intended for "low-cost" stereo receiver applications.—"Linear Databook III." Reprinted with permission of National Semiconductor.

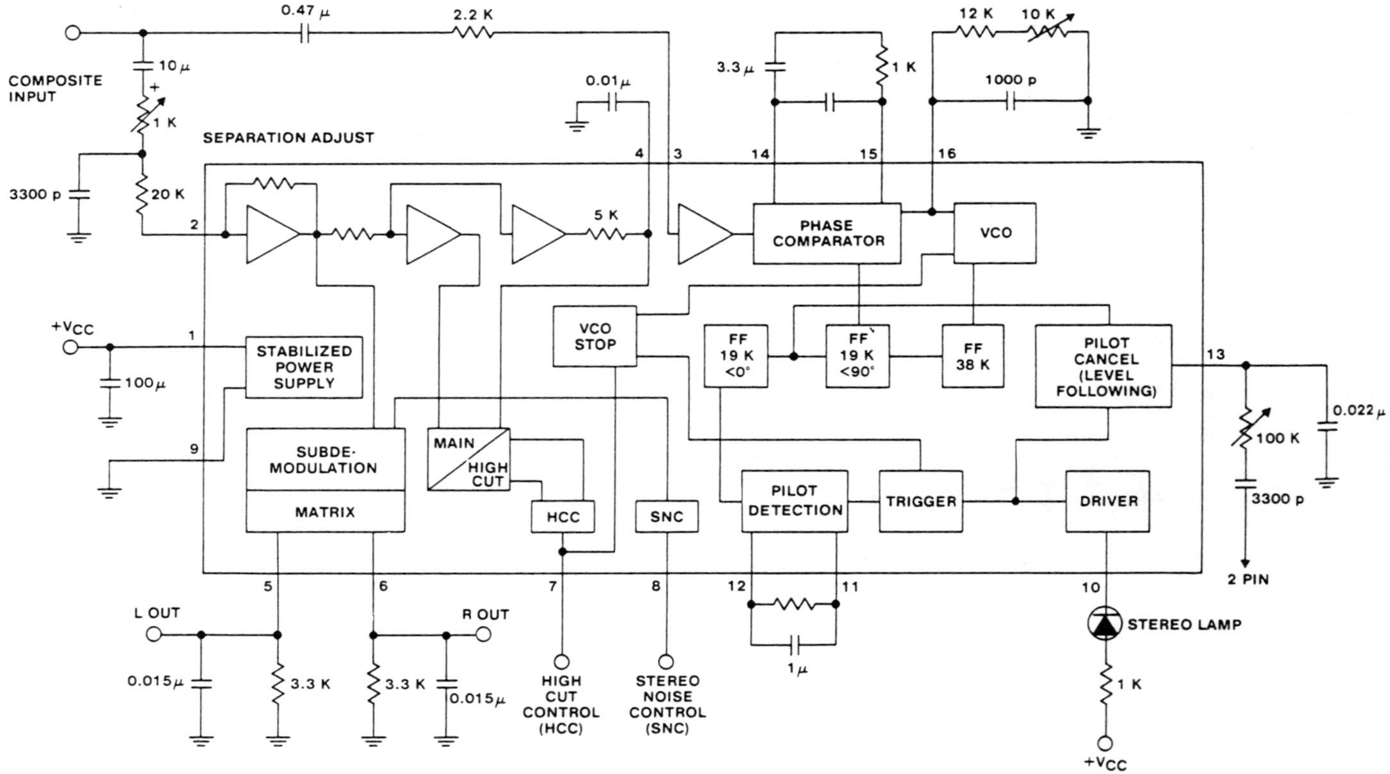

PLL FM Multiplex Stereo Demodulator. Here is an inexpensive and simple way to a high-quality FM multiplex demodulator. All of the complexity is handled on the chip, proper. The CA3257 is a complete PLL FM multiplex stereo demodulator in a single package. Only a few external components are required to put this circuit on line. This IC contains a pilot canceler, stereo noise controller, high-cut controller, automatic selection of stereo/mono, and VCO oscillator stop. A wide operating voltage range spans 6.5 to 14 V, with 12 V DC being optimal.—"RCA Solid State Databook—Integrated Circuits for Linear Applications." Courtesy of Harris Semiconductor.

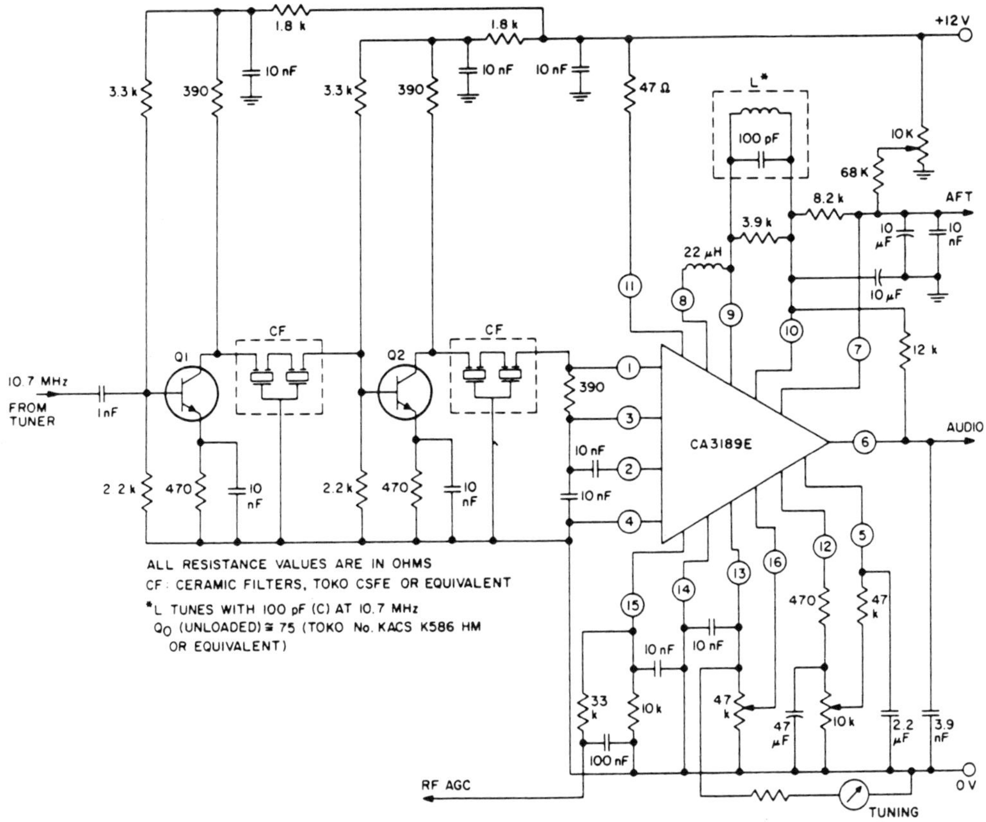

Complete FM IF System for High-Quality Receivers. For those who want to go to the trouble of building their own FM IF system, one with truly commercial quality specifications, the CA3189E FM IF system IC is a convenient way to go. This circuit offers everything anyone would expect in a high-priced commercial version, including attachment points for a tuning meter and two ceramic filters. While the circuit may be constructed on a piece of vector board, it would probably be best to etch a printed circuit board for this particular application. The circuit operates from a single-polarity 12-V power source, although this chip could also be operated at 15 V DC.—"RCA Solid State Databook—Integrated Circuits for Linear Applications." Courtesy of Harris Semiconductor.

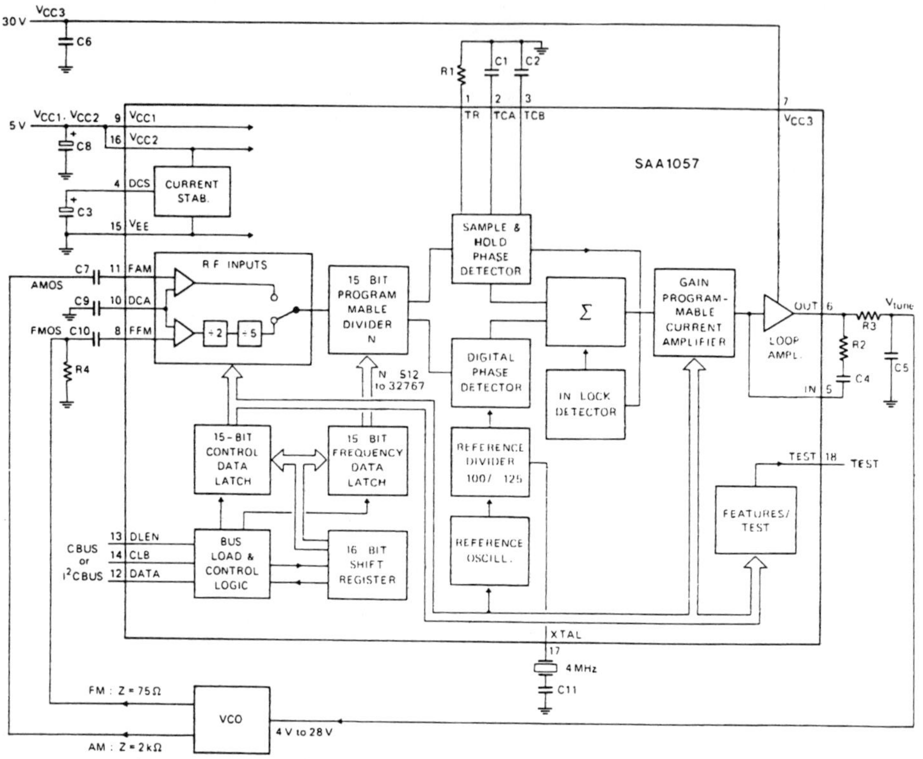

REF	FUNCTION	VALUE
R_1	Defines the current in the analog phase detector	180Ω
R_2	Loop filter resistor (value depends on V_{CO})	18kΩ
R_3	Low-pass filter resistor (value depends on V_{CO})	100Ω min 10kΩ typ
R_4	Matching resistor for 75Ω FM input	180Ω
C_1	Sample capacitor (low leakage type)	2.2nF typ
C_2	Hold capacitor (low leakage type)	10nF typ
C_3	Decoupling of internal reference voltage	47μF
C_4	Loop filter capacitor (value depends on V_{CO})	330nF typ
C_5	Low-pass filter capacitor, normally located in the tuner (value depends on loop frequency)	100nF typ
C_6	Power supply filtering	100nF
C_7	DC blocking	1nF
C_8	Power supply filtering	100μF
C_9	Decoupling of RF input stages	10nF
C_{10}	DC blocking	11nF
C_{11}	Series capacitor for crystal (value depends on crystal)	33pF

AM/FM Frequency Synthesizer. This complete synthesizer circuit uses the Signetics SAA1057 phase-locked loop radio tuning IC which has many applications in receiver circuits. The IC is so complete and versatile that very few discrete components are required for the complete synthesizer operation. In FM mode, this circuit provides a tuning range of from 87.5 to 108 MHz in steps of 10 or 12.5 kHz. It uses a 10-MHz intermediate frequency and a 32-kHz reference frequency, and it requires less than 400 msec to tune across the entire band. The separate component value chart also indicates the function of each discrete capacitor or resistor.—"Linear Data Manual Volume 1: Communications." Courtesy of Signetics Company, a division of North American Philips Corporation.

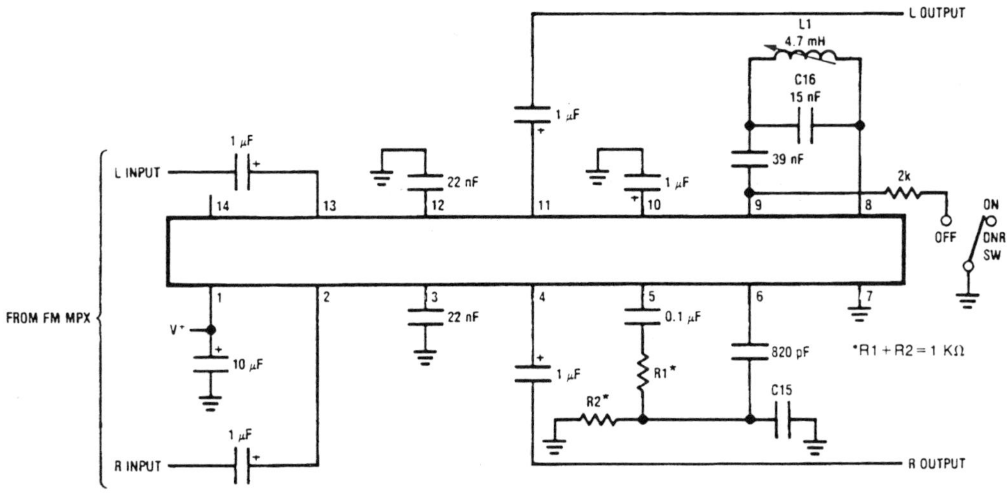

FM Stereo Dynamic Noise Reduction System. The LM832 is a dynamic noise reduction system contained on a single chip. It is designed for use with audio playback systems in general and does not require encoded source material. This means that the chip is compatible with virtually all prerecorded tapes and broadcasts. This chip is optimized for low-voltage operation with input levels around 30 mV. It will operate over a wide supply range starting at 1.5 V DC, with 9 V being the maximum. When using the chip with FM stereo as the audio source, it is necessary to eliminate the ultrasonic frequencies that will accompany the audio. If the radio has a multiplex filter to remove these ultrasonics, there will be no problem. This filtering can be done at the output of the demodulator before the DNR system or in the DNR system control path. Standard audio multiplex filters are available for use at the output of the demodulator. Coil L1 is a 4.7-mH unit that must be tuned to 19 kHz, this being the FM pilot frequency.—"Linear Databook III." Reprinted with permission of National Semiconductor.

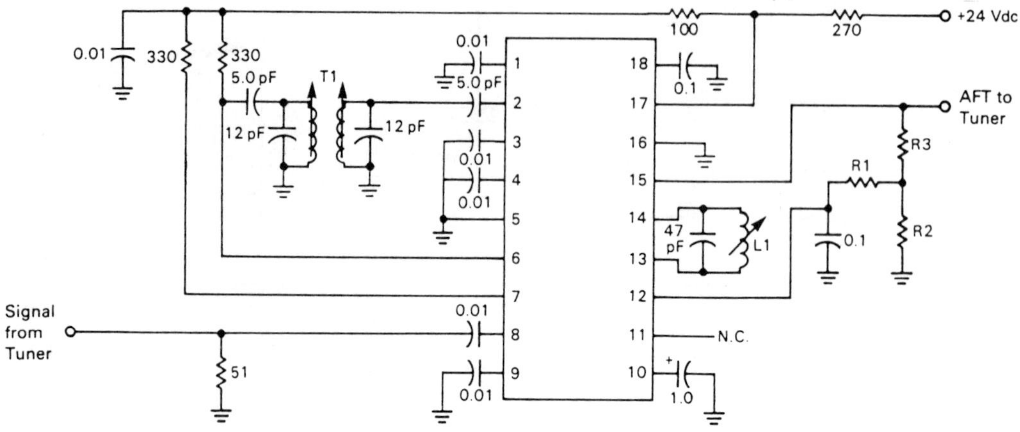

Channel 3 Component Values

T1 — Made from two coils positioned side by side, without shields, on 0.38″ centers. Coils are COILCRAFT part no. T7-142 (violet 7-1/2 turns), each with its own slug, Carbonal E, adjusted to 63 MHz (≈0.4 μH). This should give a slightly over-coupled response. A sheild to surround the coils may be required.

L1 — COILCRAFT UNI-7/150 (blue 6-1/2 turns) or UNI-10/144 (green 5-1/2 turns) shielded, adjusted to 61.25 MHz (≈ 0.14 μH)

R1 and R2 — Adjust for nominal tuning voltage

R3 — Chosen for tuning voltage swing required. Note that Pin 15 can source or sink 300 μA (typ) at the extremes of control range

Automatic Frequency Tuner. This AFT circuit uses the MC13010P parallel sound IF and AFT IC in a stand-alone application. The chip contains the entire active system to provide a tuner with this automatic control. This device performs AM detection at the intercarrier sound output. Therefore, AM-modulated digital data may be recovered. This function may be useful in cable systems when digital coding is employed. T1 is made from two coils positioned side by side without shields on 0.38-in. centers. The coils are COILCRAFT Part No. T7-142 (violet 7.5 turns), each with its own slug, Carbonal E, adjusted to 63 MHz. This will equate to approximately 0.4 μH and should give a slightly over-coupled response. A shield may eventually be required around the coils. Inductor L1 is a 0.14-μH unit. As for resistors R1 and R2, they must be adjusted for nominal tuning voltage. R3 is chosen for the tuning voltage swing required. Note that pin 15 can source or sink 300 μA at the extremes of the control range.—"Linear and Interface Integrated Circuits." Copyright of Motorola, Inc. Used by permission.

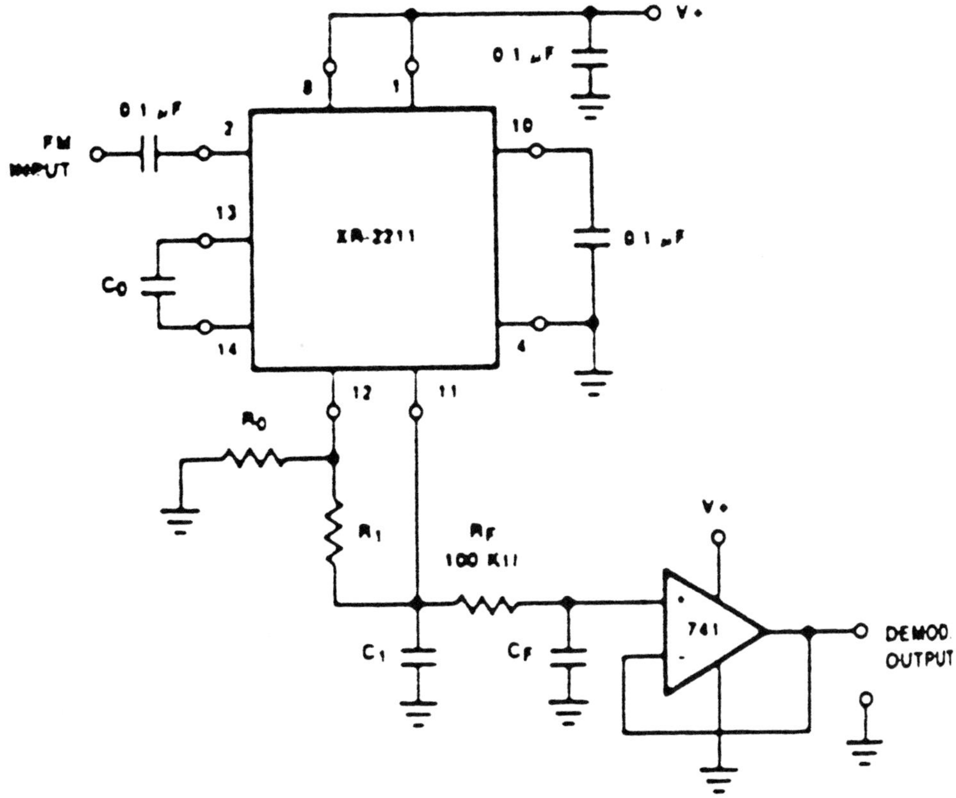

Linear FM Detector. This circuit uses the EXAR 2211 FSK demodulator/tone decoder IC as an FM detector for a wide range of analog communications and telemetry applications. The demodulated output is taken from the loop phase detector output (pin 11), through a post-detection filter made from R_f and C_f and through an external buffer amplifier. This latter component is necessary because of the high-impedance output at pin 11.—"EXAR Databook." Reproduced with the express authorization of EXAR Corporation.

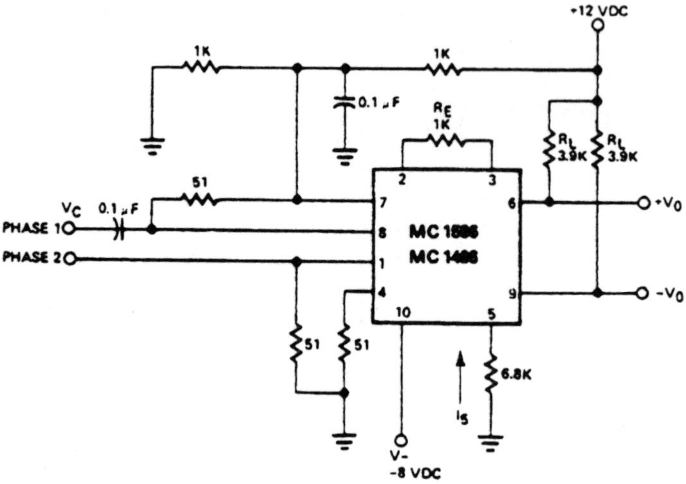

Phase Detector. Using the Signetics MC1496 balanced-modulator IC, the output of this detector circuit contains a term related to the cosine of the phase angle. Two signals of equal frequency are applied to the inputs. These frequencies are then multiplied together, thus producing the sum and difference frequencies. Equal frequencies cause the difference component to become DC, while the undesired sum component is filtered out.—"Linear Data Manual Volume 1: Communications." Courtesy of Signetics Company, a division of North American Philips Corporation.

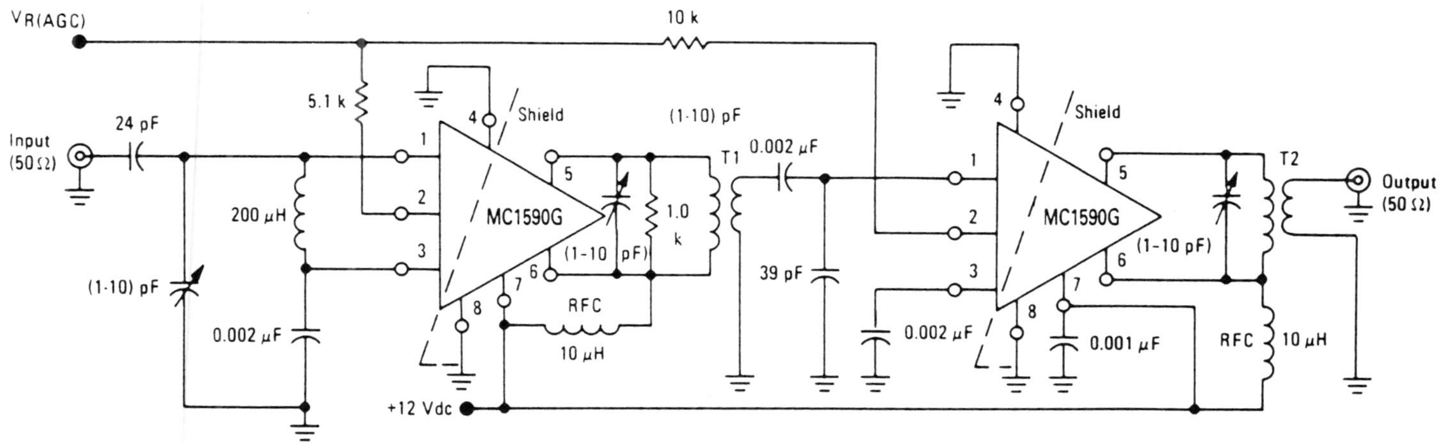

T1: Primary Winding = 15 Turns, #22 AWG Wire, 1/4" ID Air Core
Secondary Winding = 4 Turns, #22 AWG Wire,
Coefficient of Coupling ≈ 1.0

T2: Primary Winding = 10 Turns, #22 AWG Wire, 1/4" ID Air Core
Secondary Winding = 2 Turns, #22 AWG Wire,
Coefficient of Coupling ≈ 1.0

60-Megahertz IF Amplifier. This two-stage amplifier uses two Motorola MC1590G wideband amplifier ICs in a circuit that delivers a gain of approximately 80 dB at 60 MHz. It exhibits a bandwidth of 1.5 MHz. Note the shielding designations in this schematic. Adequate shielding must be supplied to separate the two amplifier stages in order to prevent unwanted feedback.—"Linear and Interface Integrated Circuits." Copyright of Motorola, Inc. Used by permission.

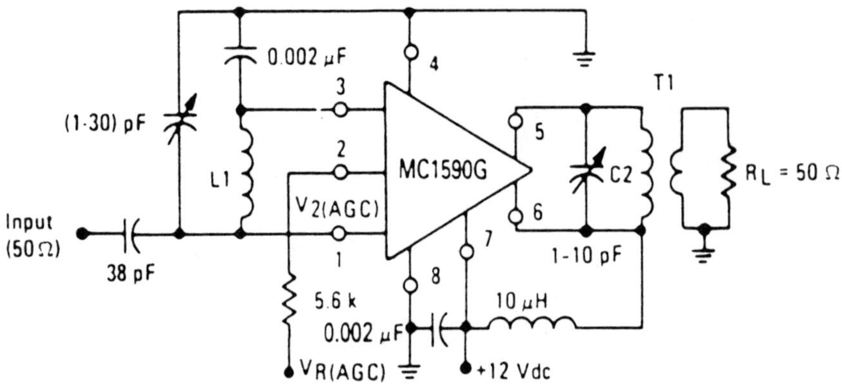

L1 = 12 Turns #22 AWG Wire on a Toroid Core,
(T37-6 Micro Metal or Equiv)
T1: Primary = 17 Turns #20 AWG Wire on a Toroid Core,
(T44-6 Micro Metal or Equiv)
Secondary = 2 Turns #20 AWG Wire

30-Megahertz Amplifier. The Motorola wideband amplifier IC is incorporated in a minimum-component, 30-MHz amplifier that may be used for many different applications including amateur radio work. This circuit will produce a power gain of approximately 50 dB and has a 1.0-MHz bandwidth.—"Linear and Interface Integrated Circuits." Copyright of Motorola, Inc. Used by permission.

Chapter 8

Video/Television Circuits

One of the best known types of linear ICs is the video amplifier. A single IC now replaces racks of discrete components that were once required to build such circuits. This chapter features a few video amplifiers but also concentrates on the multitude of specialized chips that contain whole block sections of television receivers. No company (to my knowledge) has yet designed an entire television receiver circuit on a single chip, but the day is coming. Just as chips are now available that contain 95% of an AM or FM receiver, the same will happen with television.

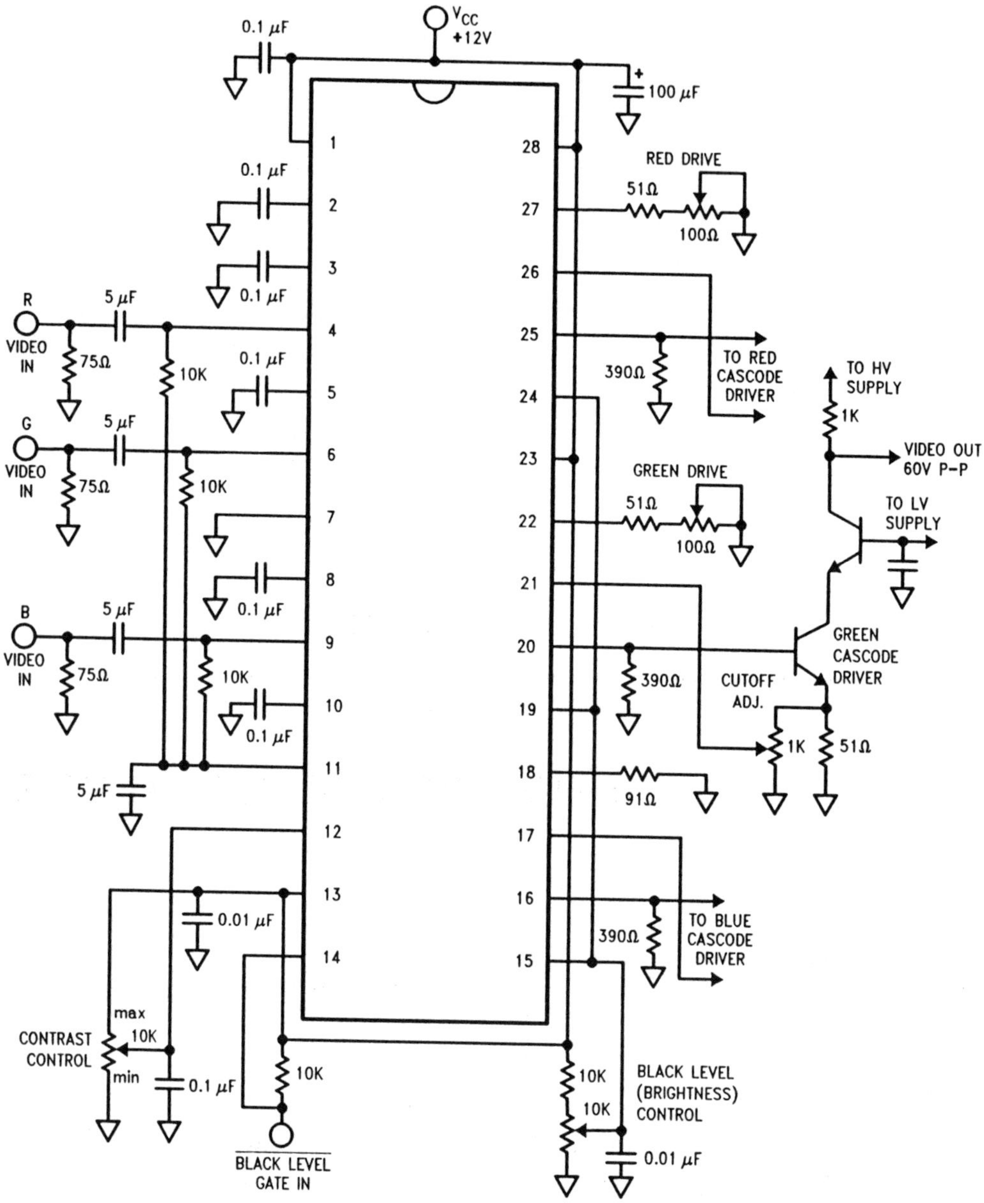

RGB Video Amplifier System. Using the National Semiconductor LM1203 wideband video amplifier system, this circuit is intended for high-resolution RGB color monitor applications. The chip contains three matched video amplifiers and three gated differential input black level clamp comparators for brightness control. There are also three matched attenuator circuits for contrast. This circuit is designed to operate from a 12-V DC source, which should be maintained. Note that this potential is very close to the 13.6-V DC absolute maximum.—"Linear Databook III." Reprinted with permission of National Semiconductor.

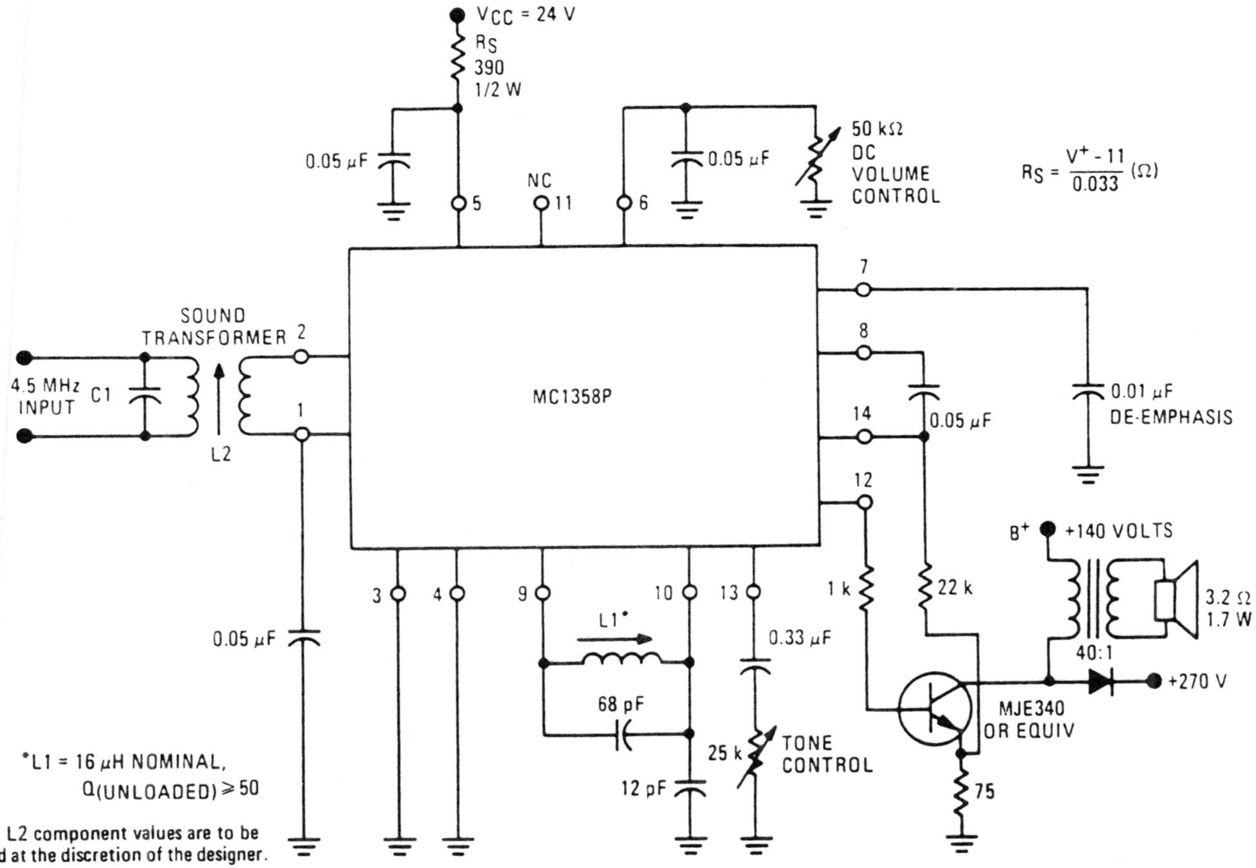

*L1 = 16 µH NOMINAL,
Q(UNLOADED) ≥ 50

C1 and L2 component values are to be selected at the discretion of the designer.

TV Sound System. This simple circuit requires one IC and very little else. It uses the Motorola MC1358P, which contains an IF amplifier, limiter, FM detector, audio driver, and electronic attenuator, all on the same chip. The 4.5-MHz input is inductively coupled to the input of the chip, while the output feeds the base of an N-P-N transistor. A tone control and DC volume control are also provided.—"Linear and Interface Integrated Circuits." Copyright of Motorola, Inc. Used by permission.

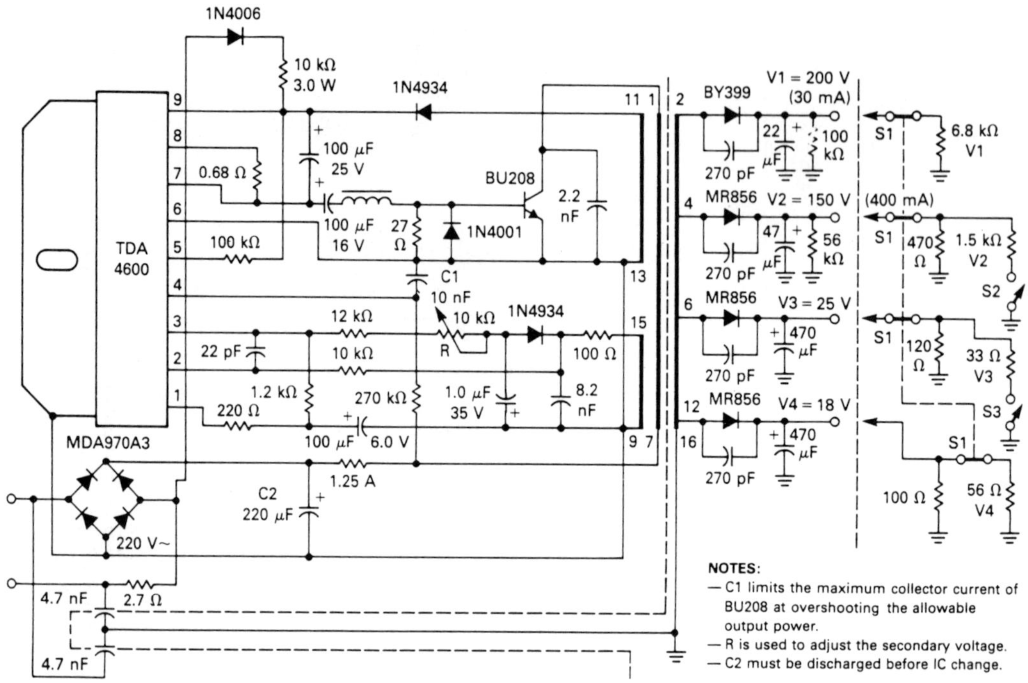

Blocking Converter for Color Television. Using the Motorola TDA4600 control IC, this circuit serves as a blocking converter for color TV sets with 30–120 W of output power and line voltages from 160 to 270 V. Only a few external components are required, and voltage stability of the various secondary voltages is achieved even with large load changes and in spite of regulation on the primary side. For line isolation and transformation to the desired secondary voltages, a transformer with a ferrite core is used.—"Linear and Interface Integrated Circuits." Copyright of Motorola, Inc. Used by permission.

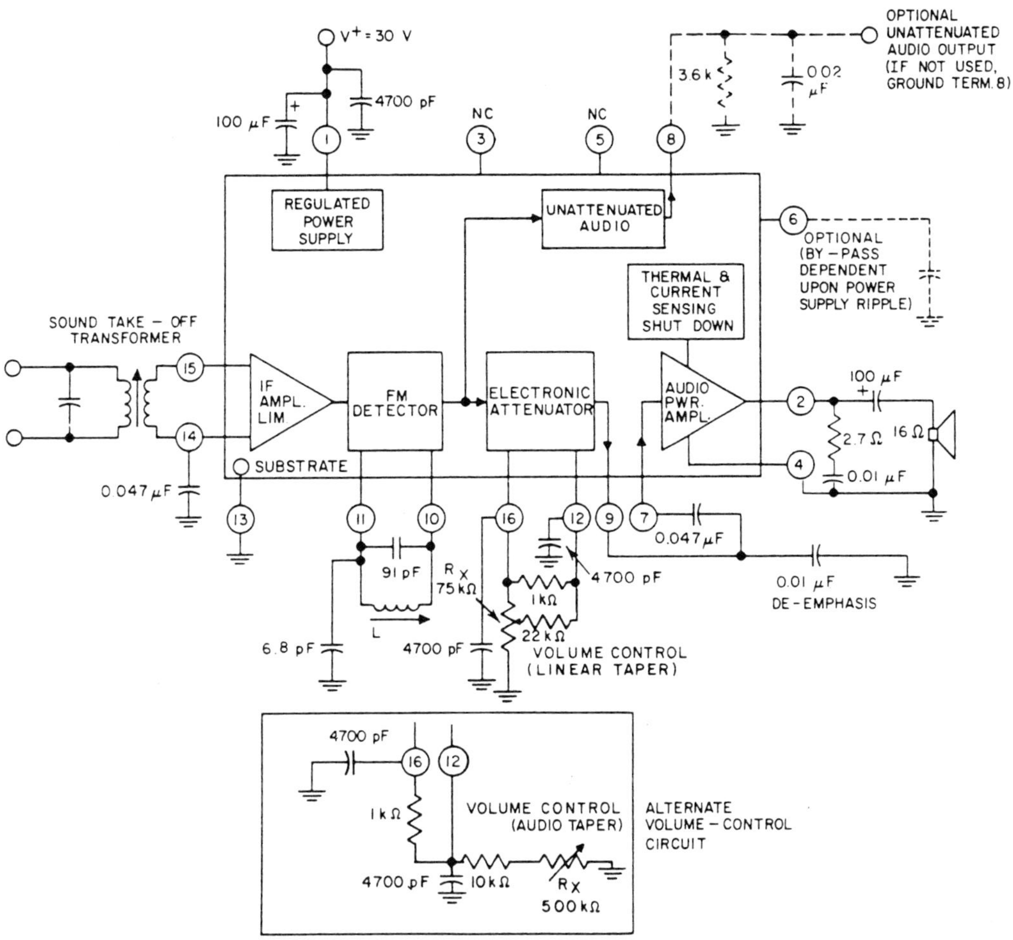

TV Sound IF and Audio. This schematic shows a block diagram of the RCA CA3134 TV sound IF and audio output subsystem IC in a typical circuit application. This IC contains a multistage IF amplifier/limiter, an FM detector, and an audio power amplifier that is designed to drive an 8- , 16- , or 32-Ω speaker. Two volume control configurations are shown. The one on the main schematic uses a linear taper potentiometer (22 kΩ), while the one in the window uses a 500-kΩ audio taper control. It is easy to see that 99% of this system is contained on the IC and very few external components are required.—"RCA Solid State Databook—Integrated Circuits for Linear Applications." Courtesy of Harris Semiconductor.

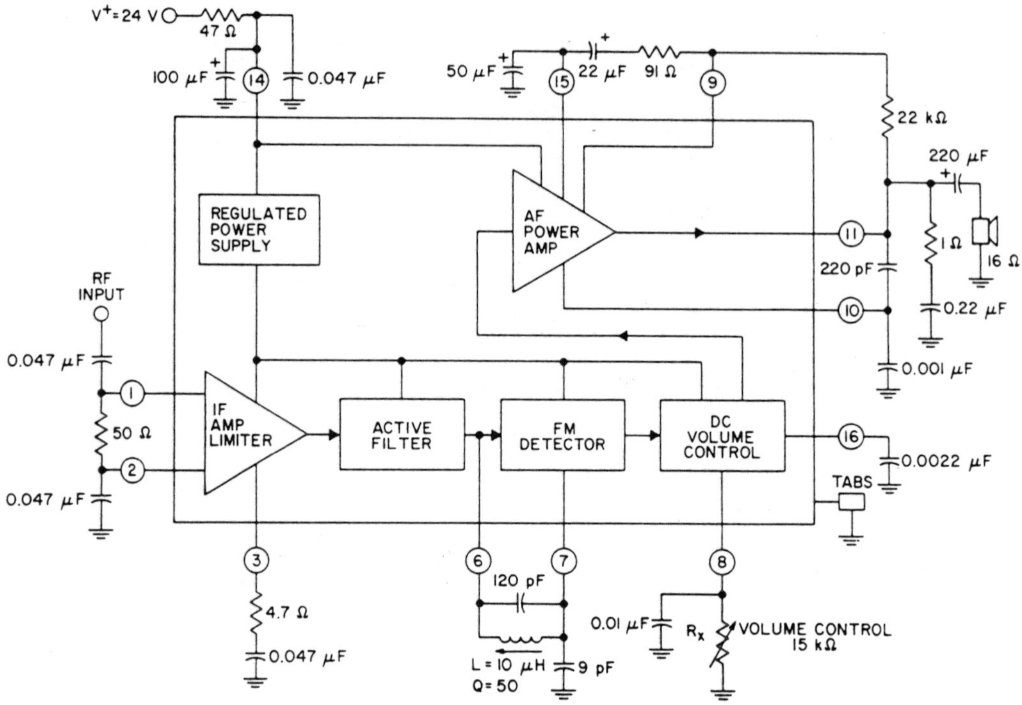

Television Sound IF and Audio Module. Using the RCA CA1190 TV sound IF and audio IC, this circuit needs only a few resistors and capacitors and a single inductor to arrive at a comprehensive television system module. The circuit is straightforward and simple due to the versatility of the IC, which offers up to a 4-W audio output. While this circuit was designed around a 24-V power supply, the IC is capable of efficient operation at voltage values as low as 12 V DC. Incidentally, this chip is electrically (and mechanically) equivalent to the industry type TDA1190Z.—"RCA Solid State Databook—Integrated Circuits for Linear Applications." Courtesy of Harris Semiconductor.

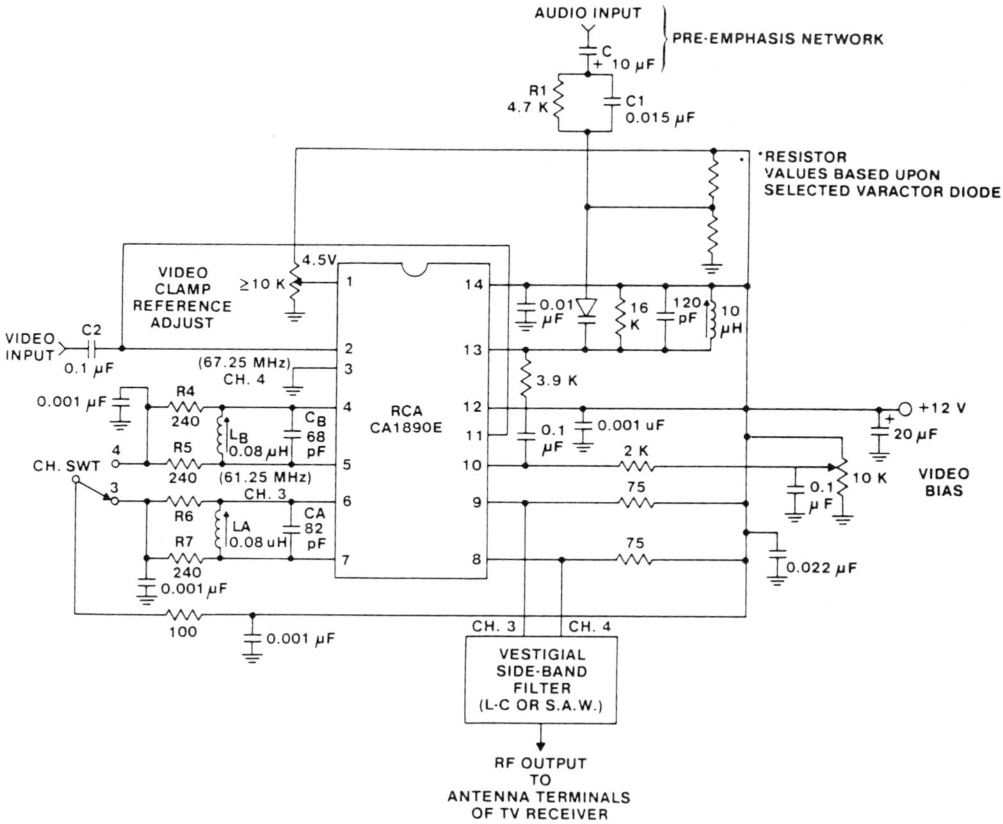

Television Video/Audio, RF Modulator. The RCA RF modulator IC forms the basis for this circuit which uses a relatively small number of external components. The purpose of this circuit is to modulate baseband audio and video signals onto one or two low-VHF TV receiver channels. In this circuit, the input video signal is AC coupled via C2. The video clamp is provided to restore/establish the DC voltage level of the AC-coupled video signal to a fixed level of the sync tips, which will help ensure proper RF modulation. Regarding the AM video oscillator portion, two channels are provided with carrier frequencies up to 110 MHz being determined by an external L–C tank circuit at pins 4 and 5 and at pins 6 and 7. In addition to the video carrier frequency inputs from either of the selected channel oscillators, the clamped video signal is input at pin 11. The video bias level and modulated sound carrier are input at pin 10. When the signal inputs are perfectly balanced, there is no RF carrier present at the output (ideally).—"RCA Solid State Databook—Integrated Circuits for Linear Applications." Courtesy of Harris Semiconductor.

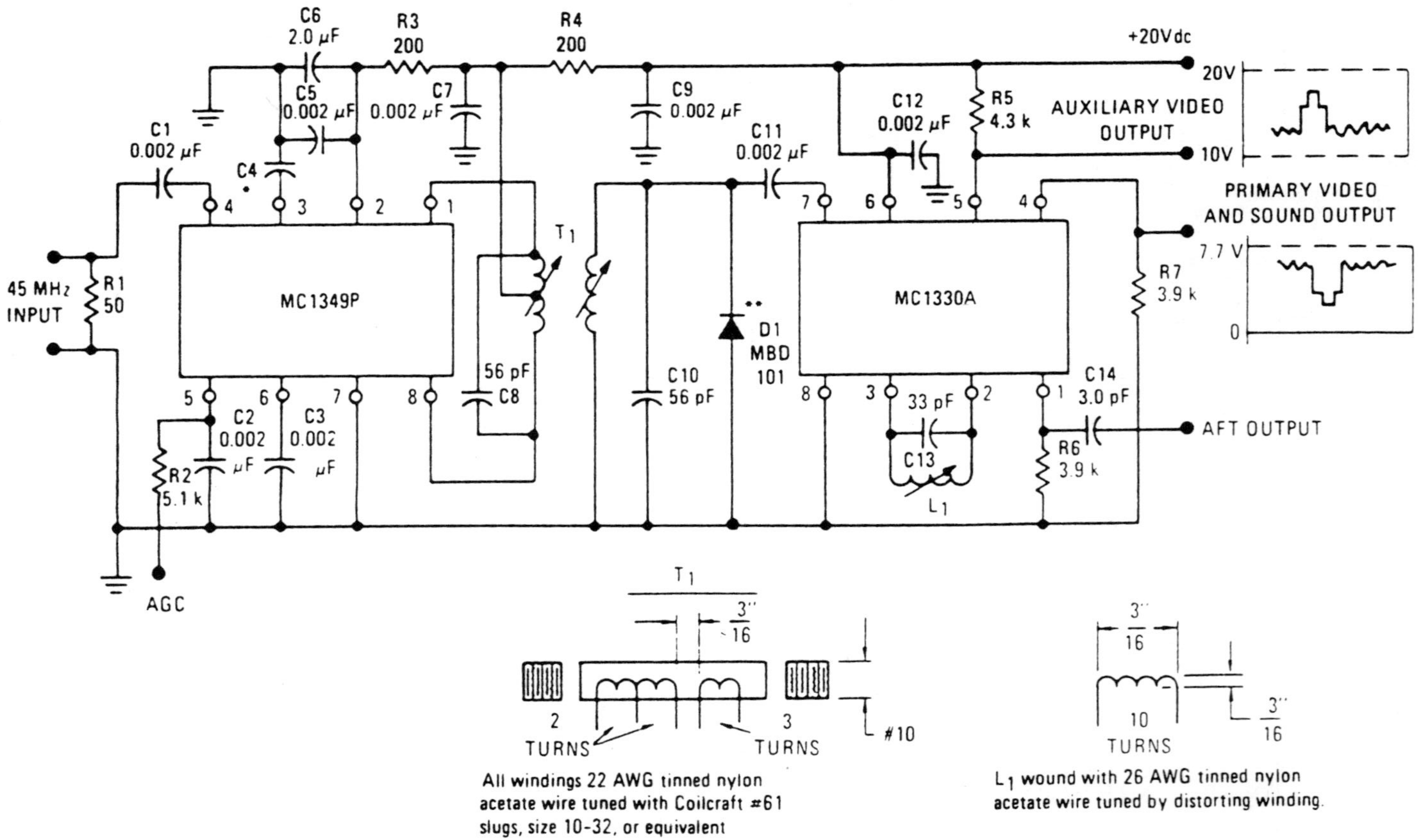

Video IF Amplifier. A very compact, high-performance IF amplifier can be constructed from this schematic. The circuit requires a bare minimum number of components and only a few alignment adjustments. It can easily be combined with normal tuners and input tuning–trapping circuitry to provide the highest quality performance. This circuit will provide a voltage gain of approximately 93 dB and can accommodate the usual low-impedance input network.—"Linear and Interface Integrated Circuits." Copyright of Motorola, Inc. Used by permission.

Chapter 9

Indicator Circuits

This is an unusual name for a category of circuits, but it is accurate. Each circuit in this section drives some sort of visual or aural reference. Here you will find VTVMs built from a single chip, linear IC thermometers, peak detectors, ammeters, battery condition indicators, and so on. Many of these will be most appropriate on the electronics test bench. Others will have very practical uses around the home or office. All can be incorporated in new equipment or in existing equipment to serve as a means to monitor operation or to test for unwanted conditions.

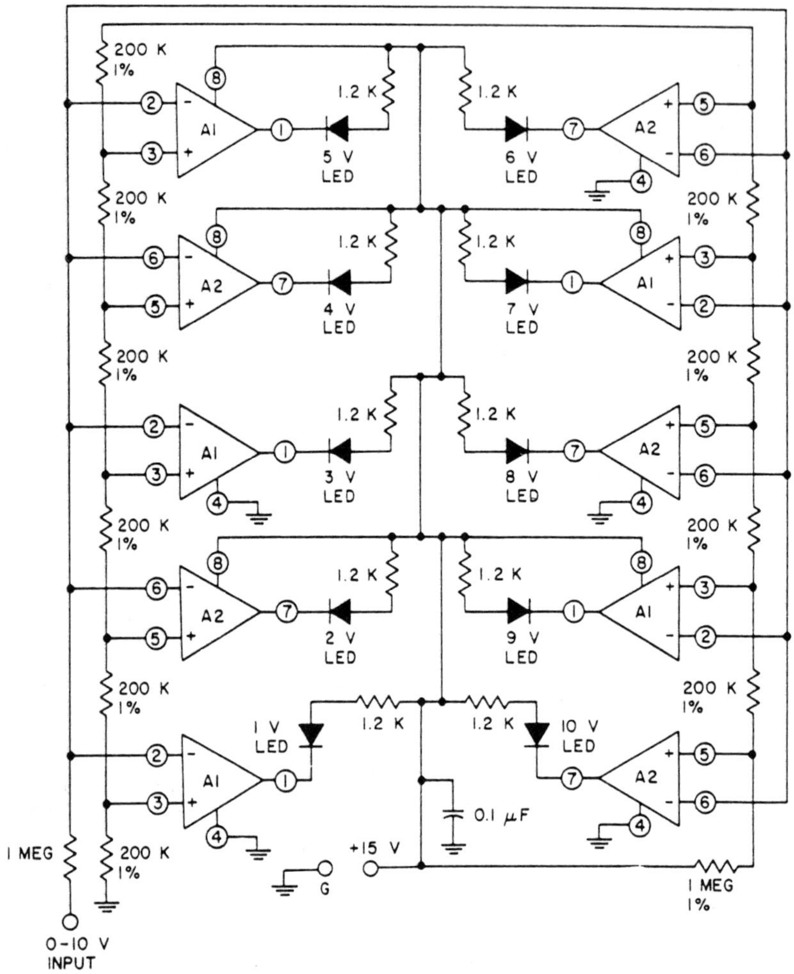

LED Bar Graph Driver. The noninverting inputs of the two CA3290 comparator circuits in this dual comparator IC package are tied to the voltage divider reference and the input signal is applied to both of the inverting inputs. The LED for a particular comparator will be switched on whenever the input voltage value reaches the same value as that of the resistor divider reference.—"RCA Solid State Databook—Integrated Circuits for Linear Applications." Courtesy of Harris Semiconductor.

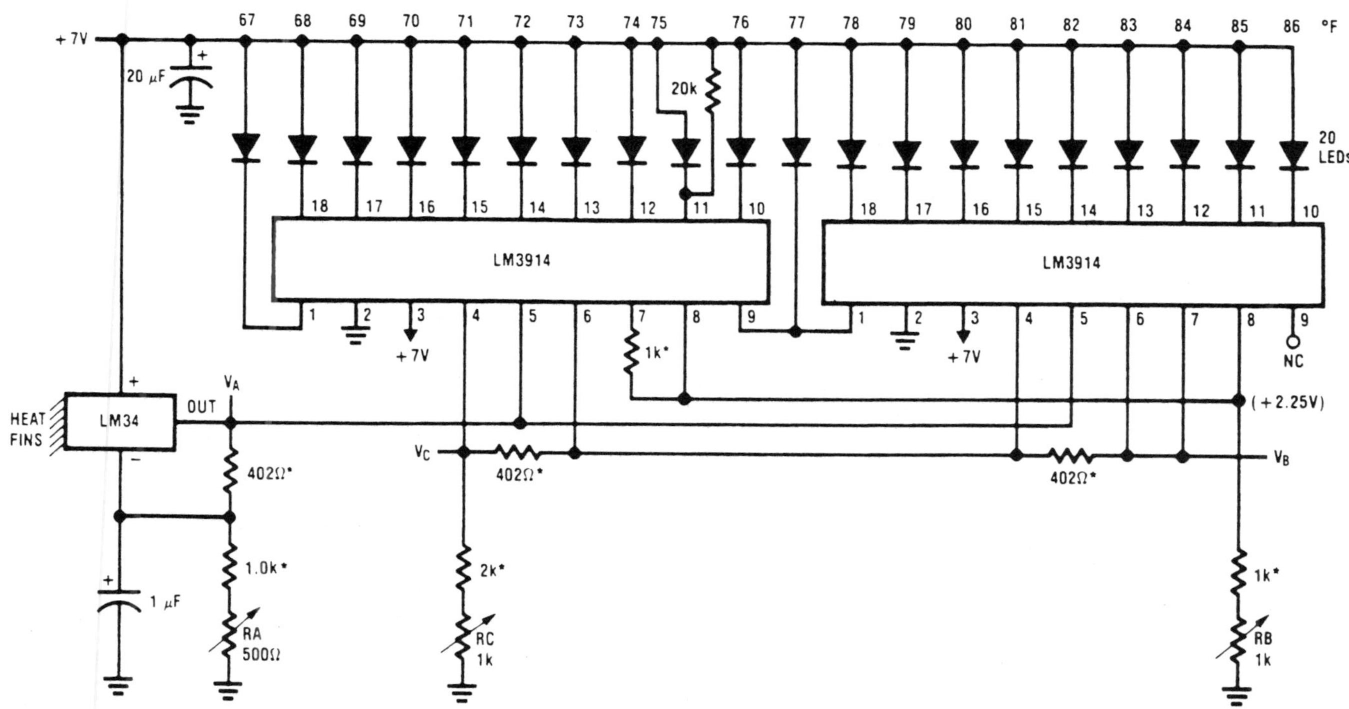

Bar-Graph Fahrenheit Temperature Display. The LM34 precision Fahrenheit temperature sensor from National Semiconductor forms the measurement basis for this simple circuit. This IC's output voltage is proportional to the Fahrenheit temperature. This is a distinct advantage over other linear temperature sensors that are usually calibrated in degrees Kelvin. This IC does not require any external calibration or trimming in order to provide outstanding accuracy which, at worst case, would be only 1.5°F over the range of −50 to 300°F. Powered by a 7-V DC supply, this circuit passes the output from the LM34 to a pair of LM3914 dot bar display ICs which drive a bank of 20 LEDs. Note that all fixed resistors are of the 1 or 2% film types for best accuracy. If such stringent accuracy requirements are unnecessary, standard carbon resistors may be combined to achieve similar results.—"Linear Databook II." Reprinted with permission of National Semiconductor.

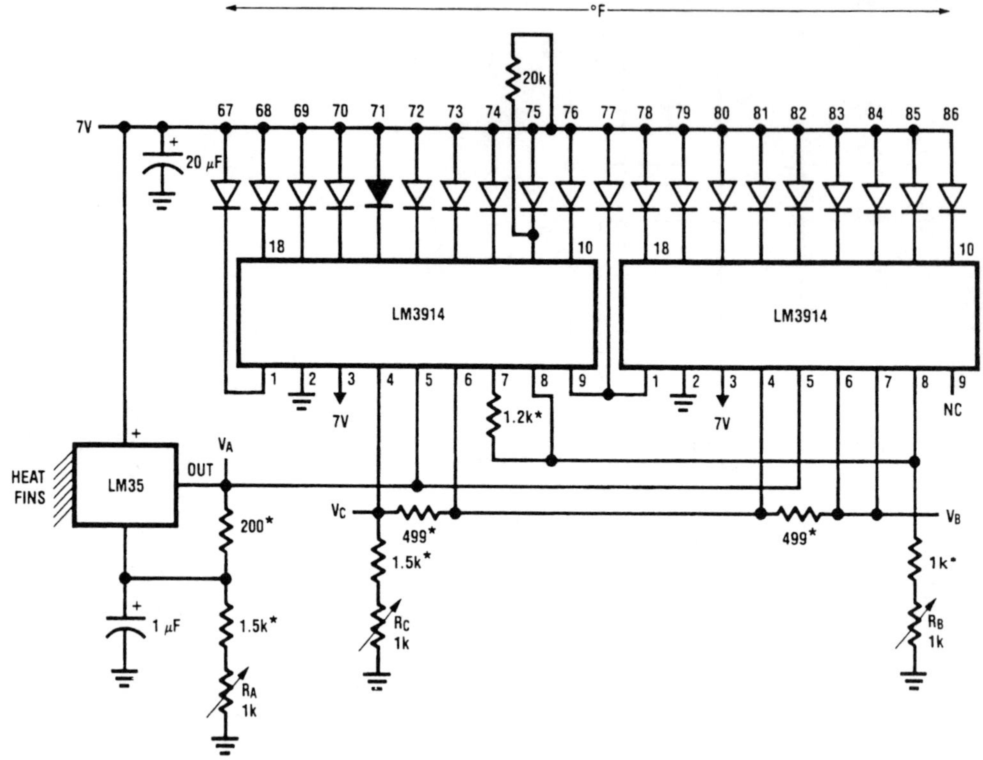

* = 1% or 2% film resistor
-Trim R_B for $V_B = 3.075V$
-Trim R_C for $V_C = 1.955V$
-Trim R_A for $V_A = 0.075V + 100mV/°C \times T_{ambient}$
-Example, $V_A = 2.275V$ at 22°C

Bar-Graph Centigrade Temperature Display. This is the same basic circuit as the previous one; however, it incorporates the LM35 precision centigrade temperature sensor which is calibrated in degrees centigrade. The rest of the circuit is basically the same, except that the fixed resistor values have been changed slightly to accommodate the different temperature scale.—"Linear Databook II." Reprinted with permission of National Semiconductor.

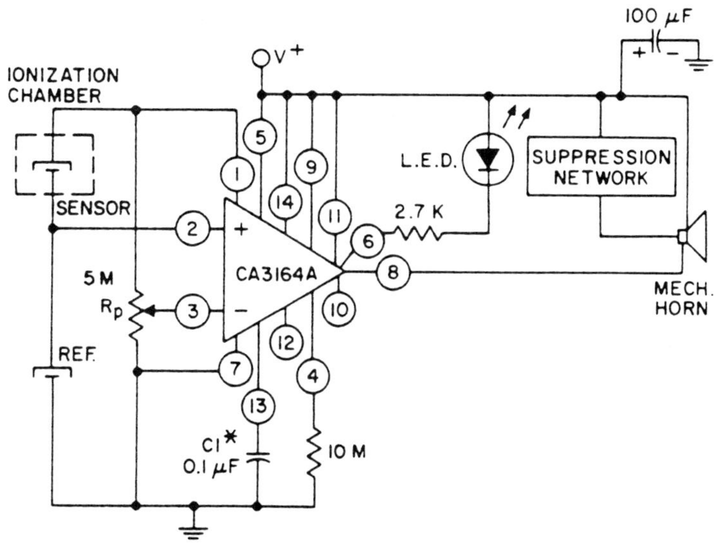

Smoke Detector. Using the CA3164A BiMOS alarm system IC and an ionization chamber, this circuit serves as an efficient smoke alarm, although it would probably be less trouble overall to simply buy one from a department store. However, designers of smoke alarm systems may find that this circuit will serve as a starting point. Capacitor C1 is used to adjust the low-battery beep level and LED blink rate. This circuit would typically operate from a 9-V battery. The IC has a limited operating voltage range of from 7 to 11 V DC. When there is no smoke in the chamber, the LED will blink every 50 sec to indicate that the circuit is operational. Naturally, the warning horn is off. When the battery is low (and there is still no smoke in the chamber), the horn will beep once every 50 sec in conjunction with the triggering of the LED. With smoke in the chamber, the horn is triggered.—"RCA Solid State Databook—Integrated Circuits for Linear Applications." Courtesy of Harris Semiconductor.

Battery Status Indicator. This tiny circuit is designed to indicate the status (good/bad) of a common 9-V battery via a visual indication provided by an LED. When the battery is connected to the input terminals, the LED will glow brightly if the voltage is above 7 V DC. Below 7 V DC there is noticeable dimming, and the LED will be extinguished at approximately 6 V DC. This is a handy little circuit that can be constructed in only a few minutes. It derives its power from the battery being tested and can be carried in a shirt pocket.—"Linear Applications Databook." Reprinted with permission of National Semiconductor.

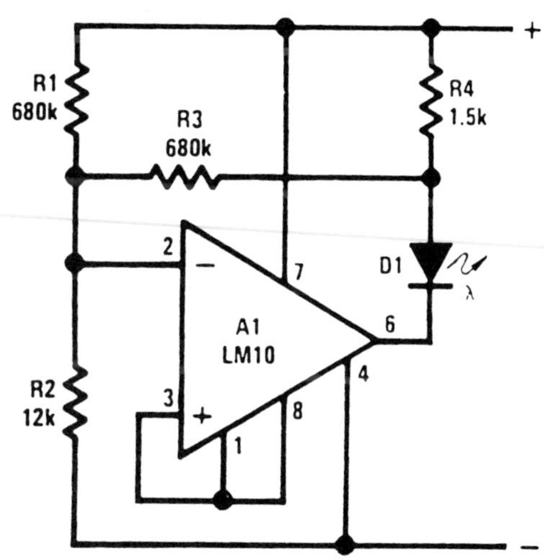

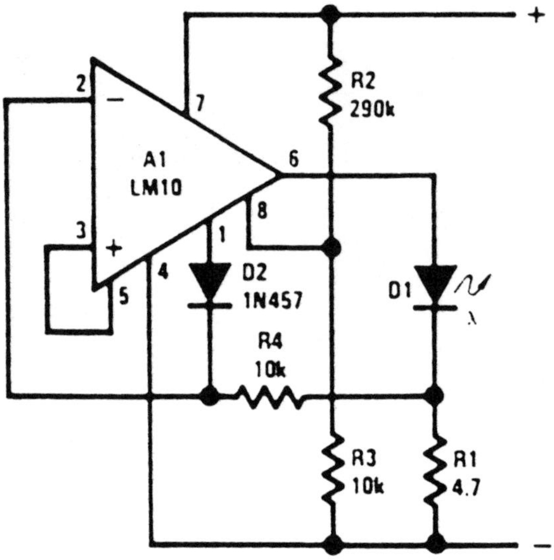

Battery Level Indicator. Very similar to the previous circuit, this one uses the LM10 op amp to regulate current through light-emitting diodes. This regulation is important in order for there to be adequate current available at minimum voltage and also to prevent excessive current from being drawn at maximum voltage. Current regulation is accomplished by using the voltage on the balance pin 5 as a reference for the op amp. In this configuration, this is controlled at approximately 23 mV, independent of temperature, by an internal regulator. The minimum threshold voltage for this circuit is limited by the bias voltage for the LEDs. Typically, this is 1.7 V for red, 2 V for green, and 2.5 V for yellow.—"Linear Applications Databook." Reprinted with permission of National Semiconductor.

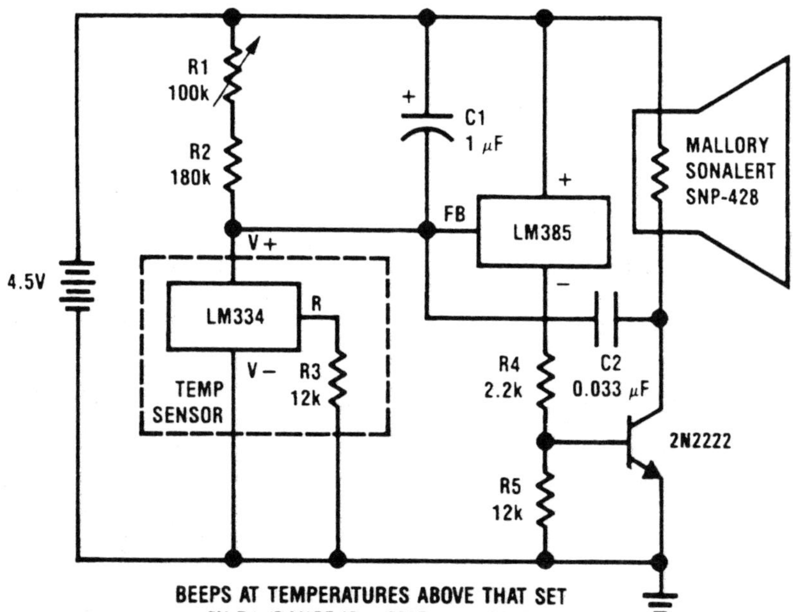

BEEPS AT TEMPERATURES ABOVE THAT SET BY R1 (RANGE IS –30°F to +120°F)

Zero-Crossing Detector. Using the LM111 voltage comparator IC, this circuit will detect zero crossing in the output of a magnetic transducer. It does this with an accuracy of a fraction of a millivolt. The magnetic pickup is connected between the two inputs of the comparator. R1 and R2 form a resistive divider which biases the inputs at 0.5 V above ground within the common mode range of the IC. The output from this circuit will directly drive DTL or TTL. The exact value of R5, the pull-up resistor, is determined by the speed required from the circuit, since it must drive any capacitive loading for positive-going output signals. An optional offset-balancing circuit using R3 and R4 is included as well.—"Linear Applications Databook." Reprinted with permission of National Semiconductor.

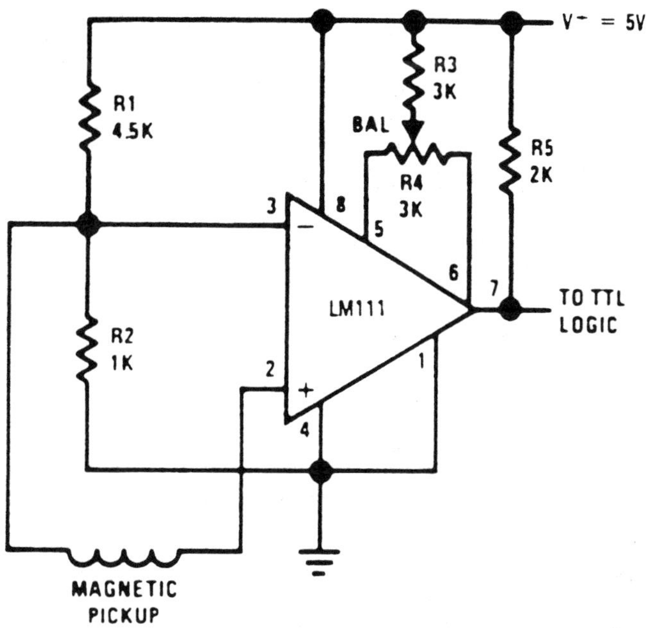

◀ **Freezer Alarm.** This is an interesting and (possibly) useful circuit designed to signal an alarm when a refrigerator, freezer, or other type of cooling device does not function properly. An LM334 adjustable current source IC is used as a temperature sensor in conjunction with an N-P-N transistor and an LM385 micropower reference. As the preset temperature maximum is exceeded (as determined by the setting of R1), the current output from the LM334 falls outside its preset range. This is detected by reference against the LM385 causing the transistor to conduct and trigger the Sonalert. A 4.5-V battery power supply is used here and may consist of three small 1.5-V DC cells. The transistor type is not critical and may be any small signal N-P-N type.—"Linear Databook II." Reprinted with permission of National Semiconductor.

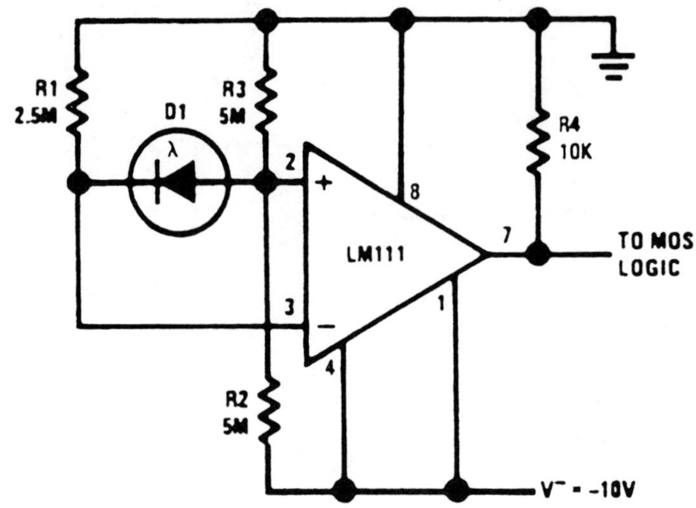

Photodiode Level Detector. Operating from a 10-V DC single-polarity supply, this circuit serves as a level detector for testing photodiodes. The output changes state when diode current reaches 1 μA. Even at this extremely low current level, comparator-contributed error is still less than 1%. If desired, higher threshold currents can be obtained by reducing the value of R1, R2, and R3 proportionally. At the switching point, the voltage across the photodiode is nearly zero. This means that its leakage current will not cause an error. The output switches between ground and −10 V DC, driving the data inputs of MOS logic directly.—"Linear Applications Databook." Reprinted with permission of National Semiconductor.

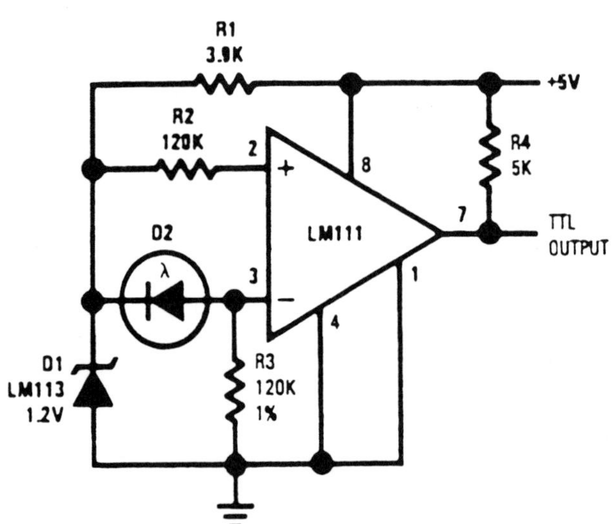

Photodiode Level Detector. This circuit is quite similar to the previous one. However, it has been adapted to work from a +5-V DC supply. The accuracy of this circuit will depend on the regulation of the supply voltage, owing to the fact that the reference is derived from the supply.—"Linear Applications Databook." Reprinted with permission of National Semiconductor.

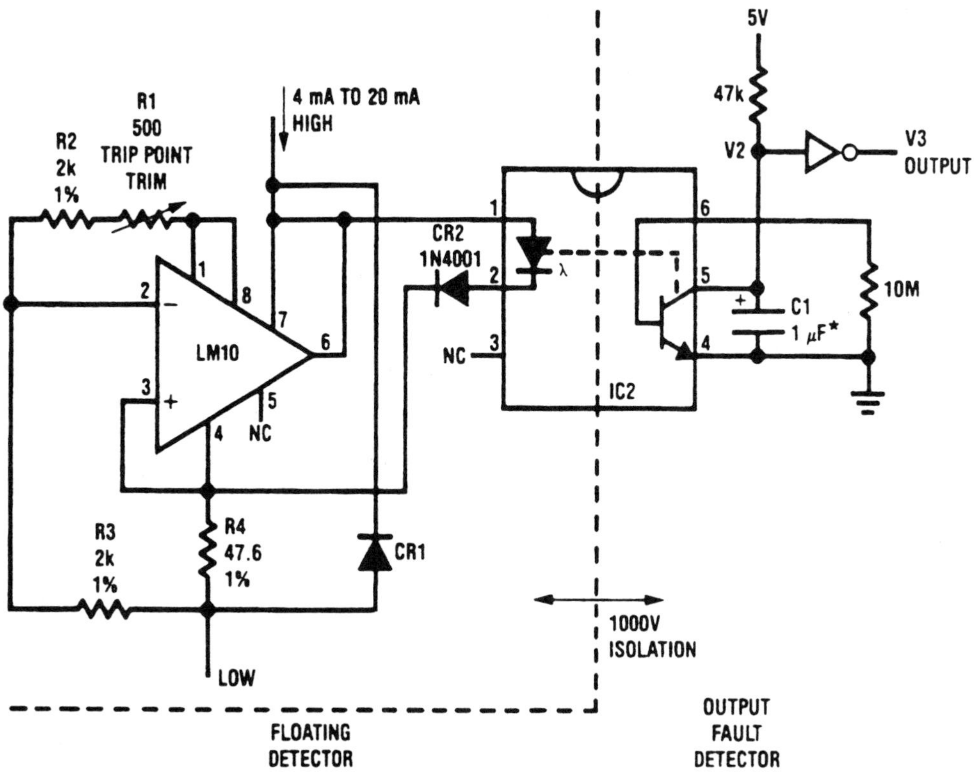

CR1 = 1N4001, optional, in case of signal reversal
LM10 = NSC LM10CLN or LM10CLH amp/reference
IC2 = 4N28 or similar, optoisolator
V2 = Normally low; high signifies fault (I < 3.7 mA)
V3 = Normally high; low signifies fault (I < 3.7 mA) (buffered output)
*C1 = 1 μF optional, to avoid false output when large AC current is superimposed on 4.0 mA.
▷o = 1/6 MM74C04 or similar, CMOS inverter

Current Loop Fault Detector. Current loops in the 4- to 20-mA range are commonly used in the process control industry. They take advantage of the fact that a remote amplifier can be powered by the same current level that it controls as its output signal, thus using a single pair of wires for both signal and power. Usually, an expensive isolation amplifier is required to detect the condition of a 4-mA signal falling out of spec without degrading the isolation of the current loop. This circuit accomplishes the same thing with simplicity and low cost. It uses the LM10 op amp and voltage reference. The stable reference is found at pins 1 and 8. Its value is 200 mV (positive) referred to pin 4. As long as the loop current is larger than 4 mA, the drop across the 47.6-Ω resistor (R4) is sufficient to pull the op amp section input (pin 2) below pin 3 and keep its output (pin 6) turned off. The loop current will flow through the LED in the optoisolator and provide a low output at pin 5 of this device. When output falls below 3.7 mA, the input at pin 2 will rise and cause the output at pin 6 to fall. This swamps all the current from the LED, causing it to extinguish.—"Linear Applications Databook." Reprinted with permission of National Semiconductor.

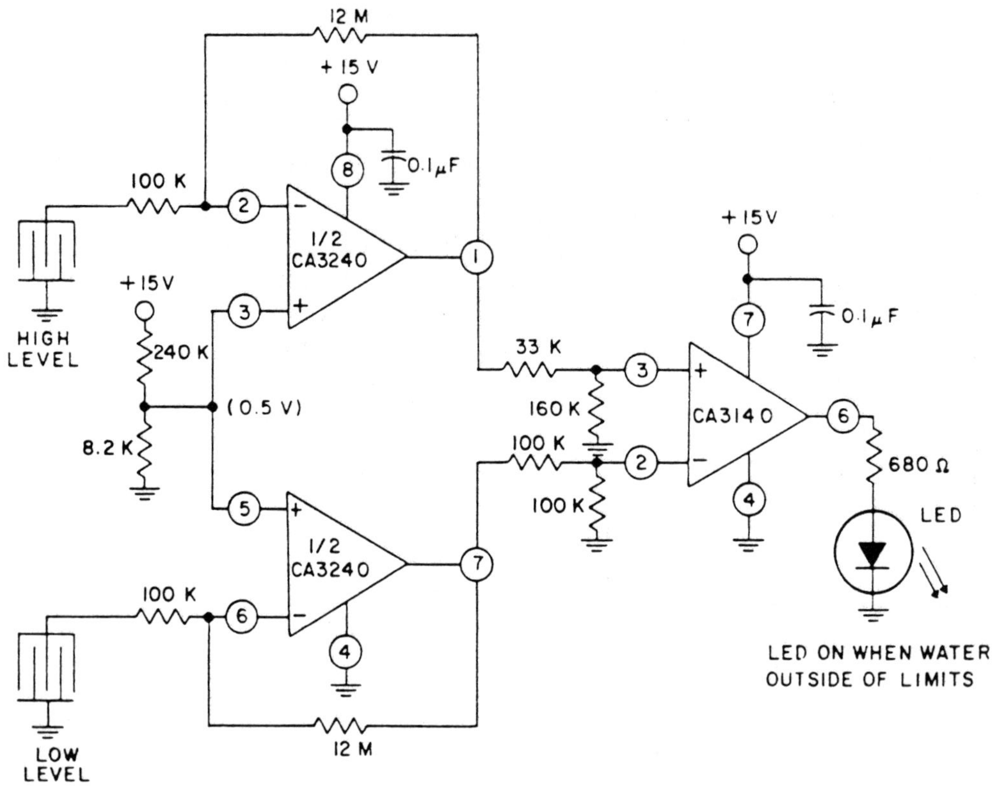

Dual Level Detector. This circuit illustrates a simple method of monitoring the levels of liquids in two separate containers. It uses the CA3240E op amp from RCA as a sensing amplifier. The circuit operates on the principle that most liquids contain enough ions in solution to sustain a small amount of current flow between two electrodes immersed in the liquid. The current, induced by the 0.5-V potential applied between two halves of a PC board grid, is converted to a voltage level by each IC amplifier. The changes in voltage for both the upper- and lower-level sensors are processed by the CA3140 op amp, which activates an LED whenever the liquid level is above the upper sensor or below the lower sensor.—"RCA Solid State Databook—Integrated Circuits for Linear Applications." Courtesy of Harris Semiconductor.

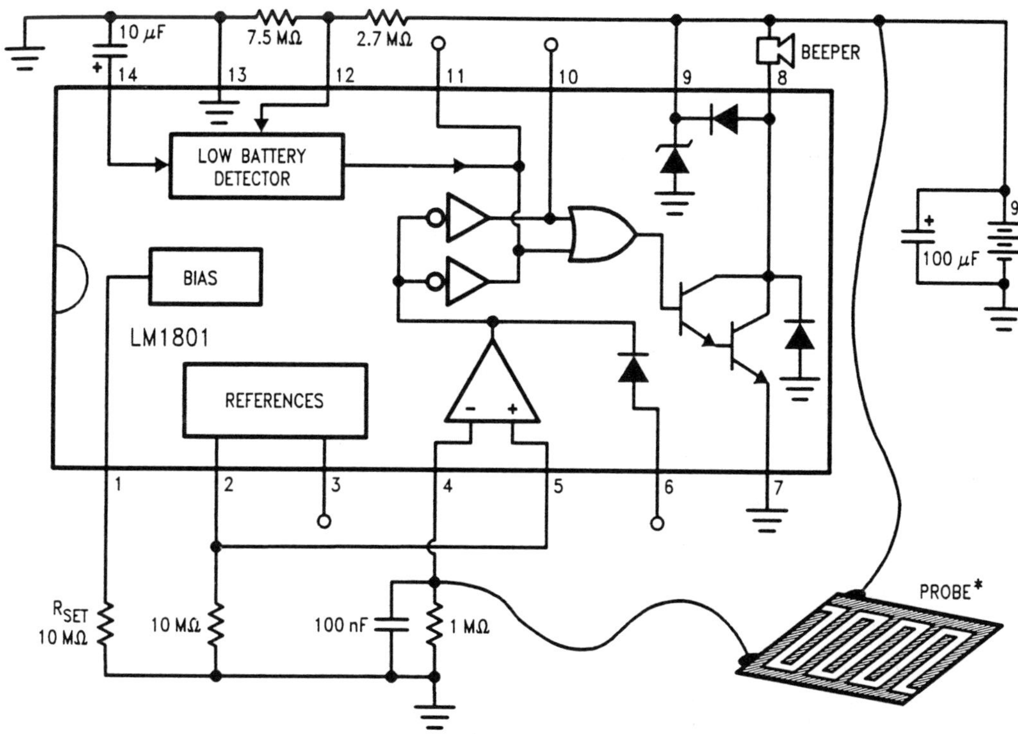

Water Leak Detector. Operated from a 9-V DC transistor radio battery, this circuit may be left untended for long periods of time without fear of discharging the battery. In this manner, it may be set and forgotten much like a modern smoke alarm. The LM1801 is an extremely low power comparator with a high-current, open-collector output stage. The typical supply current is only 7 µA, yet in its switched state the comparator can source or sink up to 0.5 A. Due to the extremely low power consumption during its "standby" state, the LM1801 can be operated from its 9-V DC battery supply for about a year. (Note: This assumes use of an alkaline-type battery.) In this circuit, a 9-V beeper or buzzer is used to sound a warning. The moisture probe is made from an etched copper-clad circuit board. The alarm sounds when the probe conductors are bridged by water droplets. This circuit is ideal for untended operation in areas where moisture build-up can cause damage.—"Linear Databook III." Reprinted with permission of National Semiconductor.

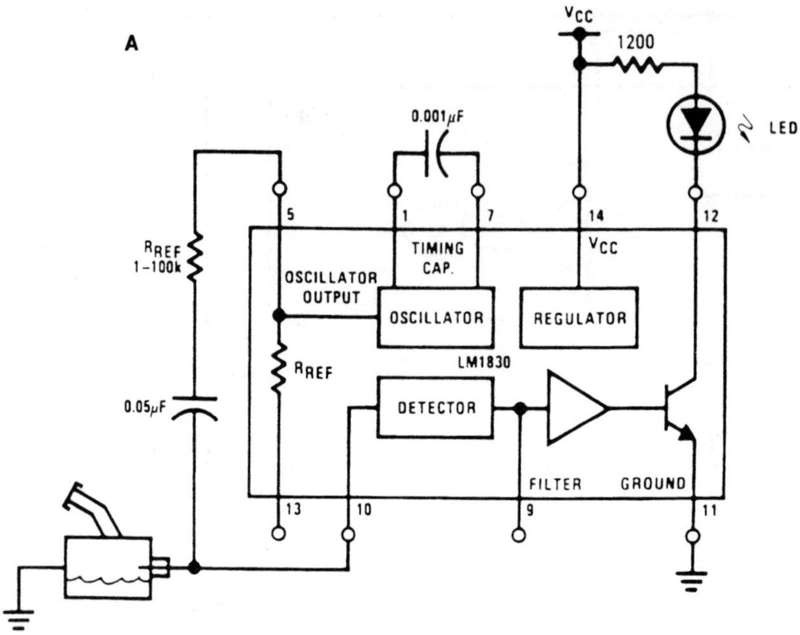

Fluid Detection Circuits. All of these circuits use a very specialized IC from National Semiconductor. The LM1830 is an IC designed specifically for fluid detection applications. An AC signal is passed through two probes within a liquid. A detector determines the presence or absence of a fluid by comparing the resistance between the fluid probes with the resistance internal to the IC. Circuit A shows the use of this IC with an external resistor. The frequency of the internal oscillator is inversely proportional to the external capacitor. Using a 0.001-μF unit as shown here, the output frequency is approximately 6 kHz. The oscillator output is available at pin 5. The internal detector is normally switched to the *on* state when the probe resistance is equal to 13 kΩ, the value of the internal resistor. However, when this high resistance value is incompatible with the probe resistance range, an external resistor may be added by coupling the probe to pin 5 through this external resistor as is the case in the first circuit. Circuit B shows an application which simply grounds the water (liquid) tank and a simple steel probe is inserted into the top of the tank. Circuit C includes a small loudspeaker to sound a warning when the internal resistance is matched. This means that other resistive transducers can also be used with this IC. Circuit D emits a 500-Hz tone when the liquid level is low. Circuit E includes an N-P-N transistor that triggers a relay or solenoid for control of higher-current warning devices. For all types of liquid level control applications the LM1830 is hard to beat. Only a few external components are required and the IC does the rest. Ideal supply voltage for these circuits is between 9 and 24 V DC.—"Linear Databook III." Reprinted with permission of National Semiconductor. (*Figure continues.*)

B

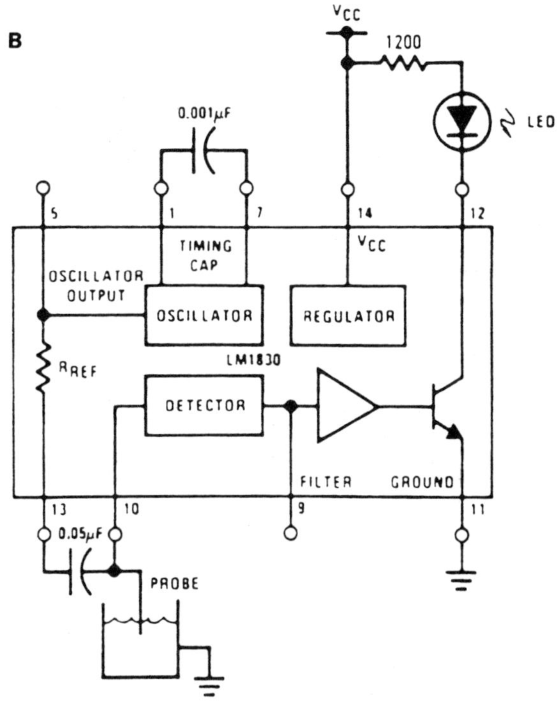
VCC
1200
LED
0.001μF
5
1
7
14
12
TIMING
CAP
VCC
OSCILLATOR
OUTPUT
OSCILLATOR
REGULATOR
LM1830
RREF
DETECTOR
FILTER
GROUND
13
10
9
11
0.05μF
PROBE

C

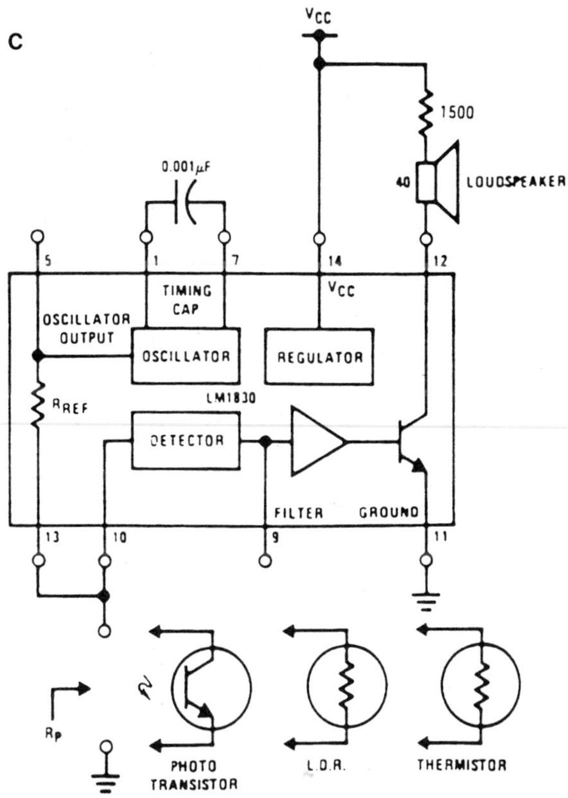
VCC
1500
40
LOUDSPEAKER
0.001μF
5
1
7
14
12
TIMING
CAP
VCC
OSCILLATOR
OUTPUT
OSCILLATOR
REGULATOR
LM1830
RREF
DETECTOR
FILTER
GROUND
13
10
9
11
RP
PHOTO
TRANSISTOR
L.D.R.
THERMISTOR

D

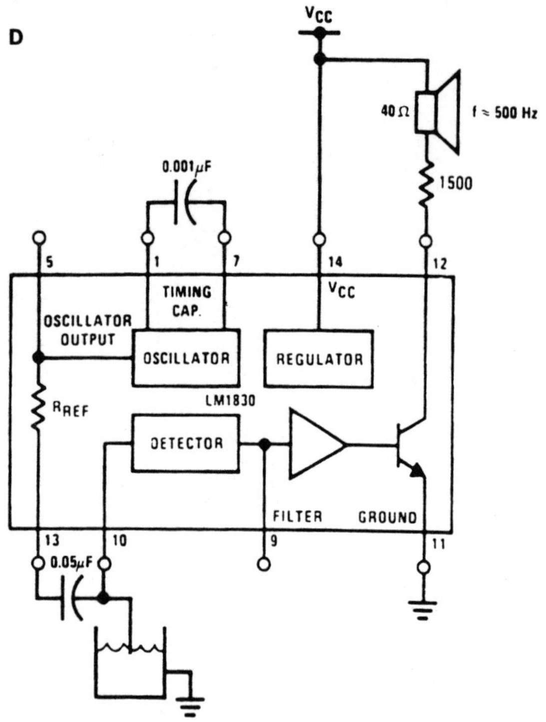

E

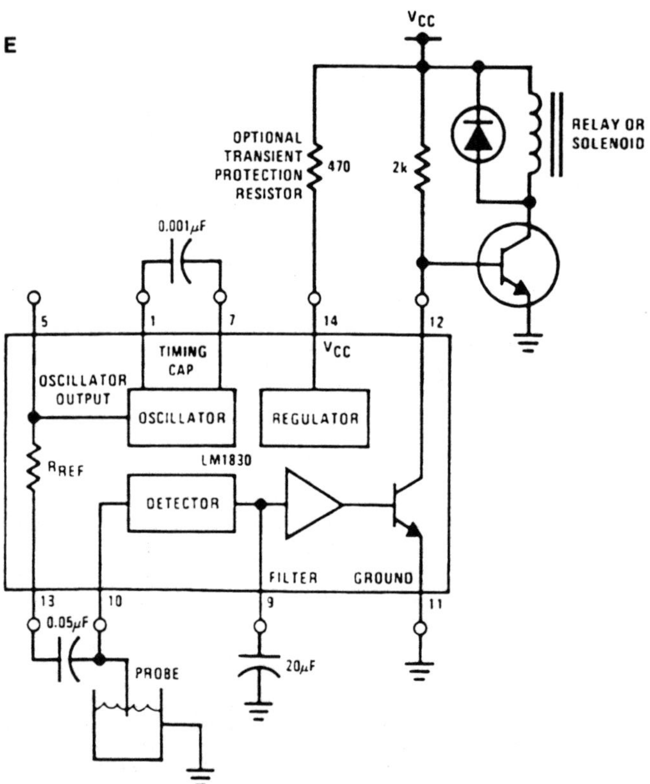

Fluid Detection Circuits *(continued)*

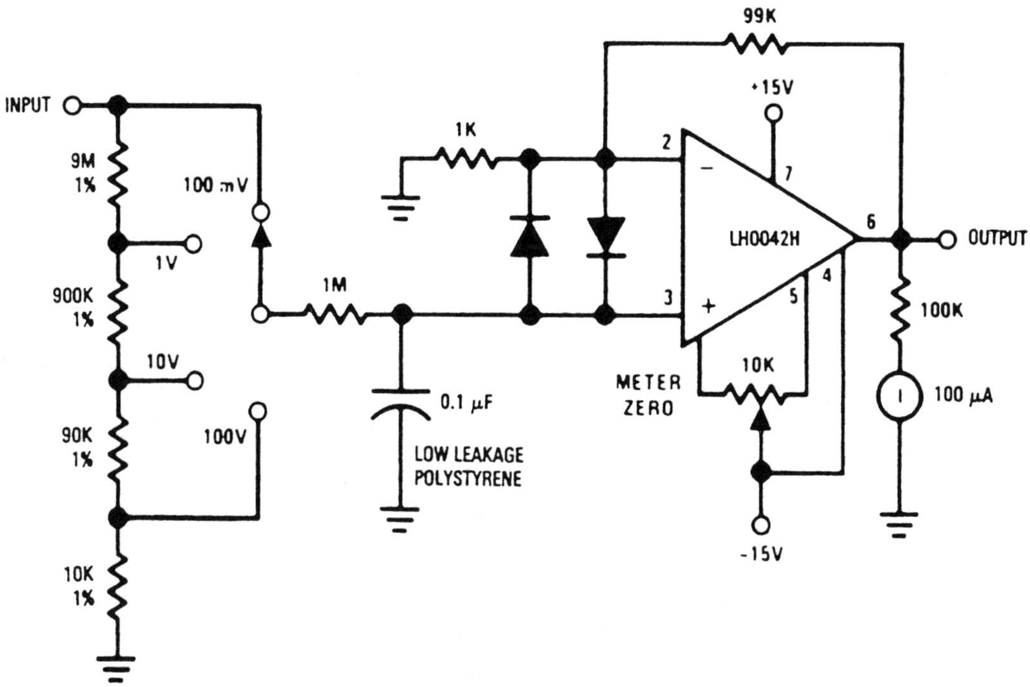

Sensitive Low-Cost "VTVM". Easy to construct, cheap, and very accurate, this circuit can be assembled on a piece of vector board and mounted directly to the back of the 100-μA meter to form a compact and handy test instrument. Naturally, the quality of the four range resistors will have a direct (and pronounced) effect on accuracy. These 1% units can be attached directly to a high-quality miniature ceramic switch that can be mounted through the meter case edge, if there is adequate space. The 10-kΩ potentiometer should also be a high-quality unit. Make certain that the capacitor is a low-leakage, polystyrene type as indicated in the schematic drawing. Any other type will not suffice.—"Linear Databook I." Reprinted with permission of National Semiconductor.

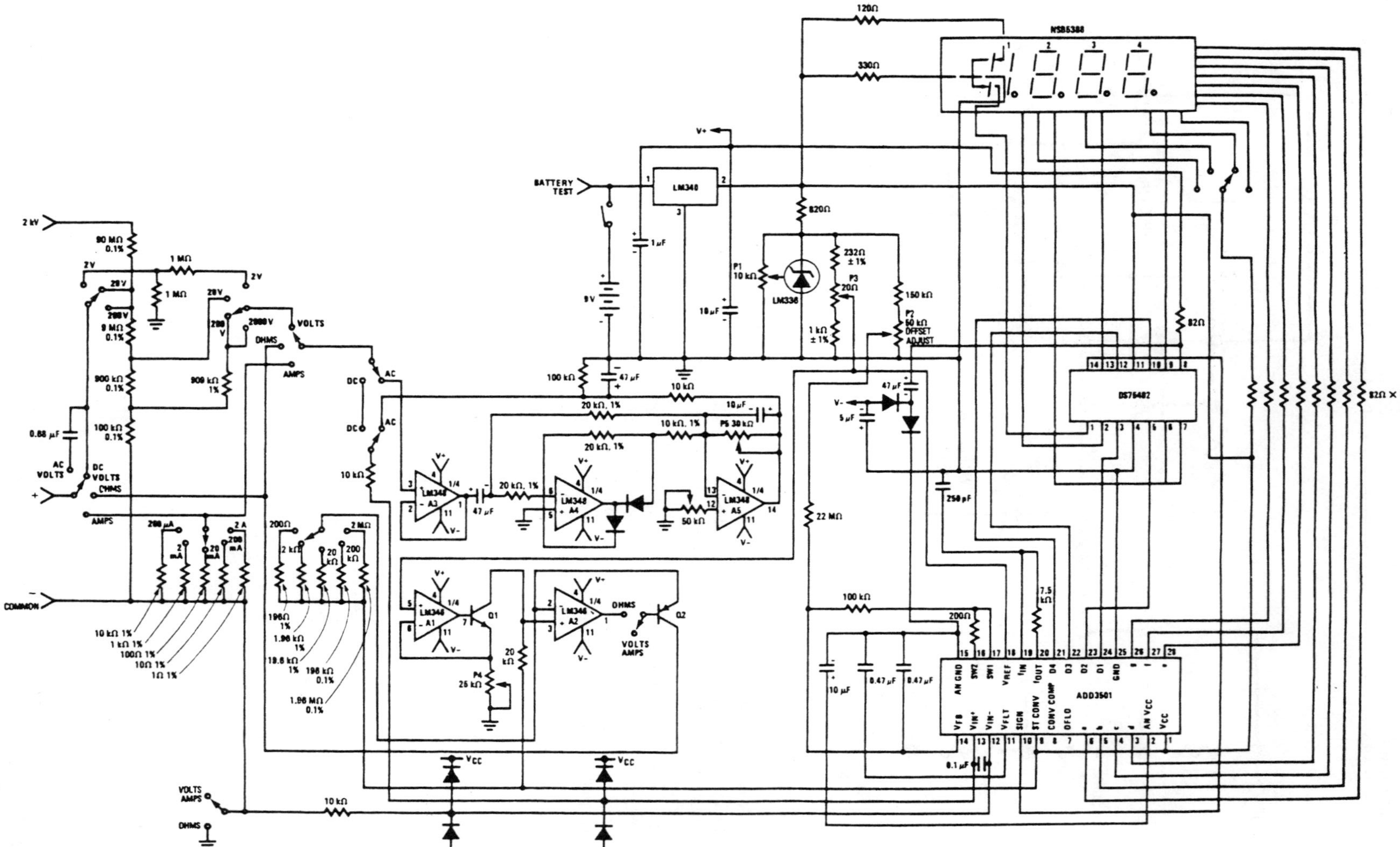
NSB5388
120Ω
330Ω
V+
BATTERY TEST
LM340
1 μF
9 V
10 μF
820Ω
P1 10 kΩ
LM336
232Ω ±1%
P3 20Ω
1 kΩ ±1%
150 kΩ
P2 50 kΩ OFFSET ADJUST
82Ω
DS75492
82Ω × 8
2 kV
90 MΩ 0.1%
1 MΩ
2V
20V
200V
2000V
9 MΩ 0.1%
900 kΩ 0.1%
100 kΩ 0.1%
900 kΩ 1%
0.88 μF
AC VOLTS
DC VOLTS
OHMS
AMPS
VOLTS
AC
DC
10 kΩ
100 kΩ
47 μF
20 kΩ, 1%
10 kΩ, 1%
P5 30 kΩ
LM348 A3
LM348 A4
LM348 A5
50 kΩ
V-
5 μF
250 pF
22 MΩ
COMMON
200 μA
2 mA
20 mA
200 mA
2 A
10 kΩ 1%
1 kΩ 1%
100Ω 1%
10Ω 1%
1Ω 1%
200Ω
2 kΩ
20 kΩ
200 kΩ
2 MΩ
196Ω 1%
1.96 kΩ 1%
19.6 kΩ 1%
196 kΩ 0.1%
1.96 MΩ 0.1%
LM346 A1
LM346 A2
Q1
Q2
20 kΩ
P4 25 kΩ
VOLTS AMPS
VCC
7.5 kΩ
200Ω
0.47 μF
0.1 μF
ADD3501
AN GND
SW2
SW1
VREF
IIN
IOUT
D4
D3
D2
D1
GND
VFB
VIN+
VIN-
VFLT
SIGN
ST CONV
CONV COMP
OFLO
AN VCC
VCC

Note 1: All V_{CC} connections should use a single V_{CC} point and all ground/analog ground connections should use a single ground/analog ground point.

Note 2: All resistors are ¼ watt unless otherwise specified.

Note 3: All capacitors are ±10%.

Note 4: All op amps have a 0.1 μF capacitor connected across the V+ and V− supplies.

Note 5: All diodes are 1N914.

DMM PERFORMANCE

Measurement Mode	Range 0.2	2	2	200	2000	Frequency Response	Accuracy	Overrange Display
DC Volts	—	V	V	V	V	—	≤ 1% F.S.	± OFLO
AC Volts	—	V_{RMS}	V_{RMS}	V_{RMS}	V_{RMS}	40 Hz to 5 kHz	≤ 1% F.S.	+ OFLO
DC Amps	mA	mA	mA	mA	mA		≤ 1% F.S.	± OFLO
AC Amps	mA_{RMS}	mA_{RMS}	mA_{RMS}	mA_{RMS}	mA_{RMS}	40 Hz to 5 kHz	≤ 1% F.S.	+ OFLO
Ohms	kΩ	kΩ	kΩ	kΩ	kΩ	—	≤ 1% F.S.	+ OFLO

TECHNICAL SPECIFICATIONS

DC VOLTS	< ±1% ACCURACY
RANGES	2V, 20V, 200V, 2 kV
INPUT IMPEDANCE	2V RANGE, >10MΩ
	20V TO 2 kV RANGE, 10MΩ
AC RMS VOLTS	< ±1% ACCURACY
RANGES	2V, 20V, 200V, 2 kV
	(40 TO 5 kHz SINEWAVE)
DC AMPS	< ±1% ACCURACY
RANGES	200 μA, 2mA, 20 mA, 200 mA, 2A
AC RMS AMPS	< ±1% ACCURACY
RANGES	200 μA, 2 mA, 20 mA, 200 mA, 2 A
OHMS	< ±1% ACCURACY
RANGES	200 Ω, 2 kΩ, 20 kΩ, 200 kΩ, 2 MΩ

Digital Multimeter. The construction of a complete digital multimeter is well within the grasp of most technicians when linear ICs designed specifically for this application are incorporated. This circuit uses National Semiconductor's ADD3501 monolithic CMOS IC, designed for use as a $3\frac{1}{2}$-digit digital voltmeter. This chip makes use of a pulse-modulation, analog-to-digital conversion scheme that operates from a 2-V reference voltage. Supporting the aforementioned chip is a DS75492 digit driver, an NSB5388 LED display, and an LM340 voltage regulator for the power supply. A 2-V reference voltage is derived from the LM336 reference diode circuitry and permits the $3\frac{1}{2}$-digit system a 1 mV/LSD resolution. For the most part, the circuit may be constructed on a moderately small piece of vector board and then installed on a small plastic or aluminum cabinet available from most electronic outlets. Alternately, you can etch your own printed circuit board, but vector board will serve nicely, especially if the on-board components are anchored with epoxy.—"Linear Applications Databook." Reprinted with permission of National Semiconductor.

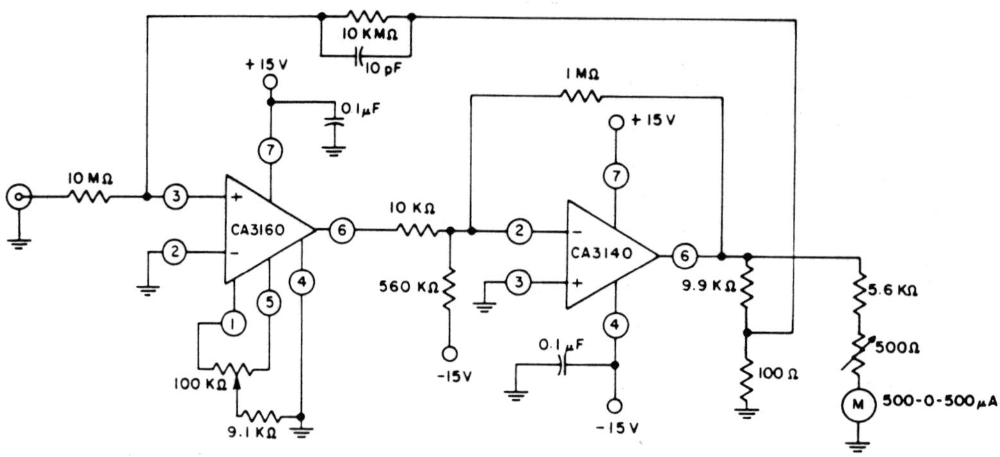

Picoammeter. This circuit provides a very sensitive picoammeter with a ±3-pA full-scale deflection. This is actually a current-to-voltage converter utilizing two operational amplifiers. By placing pins 2 and 4 of the 3160 at ground potential, its input is operated in the guarded mode. Under this condition, even slight leakage resistance between pins 3 and 2 or between pins 3 and 4 will result in zero voltage across this leakage resistance, thus substantially reducing leakage current. The 3140 op amp stage serves as a ×100 gain stage to provide the required positive and negative output swing for the meter and feedback network. A 100 : 1 voltage divider network consisting of a 9.9-kΩ resistor in series with a 100-Ω resistor sets the voltage at the resistor in series with terminal 3- to ±30-mV full-scale deflection. This value results from ±3 V appearing at the top of the voltage divider network, which is also used to drive the meter circuitry.—"RCA Solid State Databook—Integrated Circuits for Linear Applications." Courtesy of Harris Semiconductor.

Chapter 10

Automotive Circuits

This chapter features circuits that are designed specifically for automotive applications or which will certainly have applications in this field. In the purely automotive category are tachometers, speed control units, and ignition systems. Also found here are radio amplifiers and similar devices that are designed specifically to be powered by the automotive electrical system.

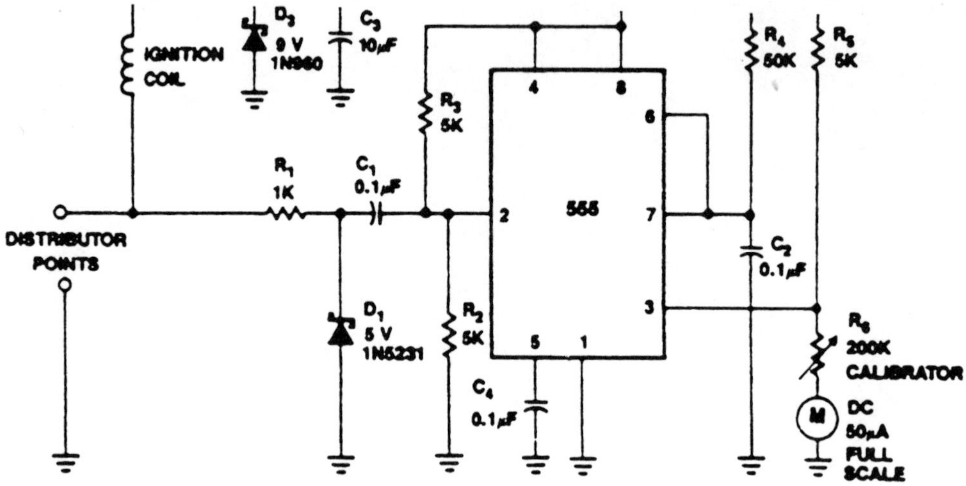

Automobile Tachometer. This simple circuit uses the old and proven 555 timer IC in a configuration that operates from the automotive electrical system with input from the distributor points. It uses a 50-μA meter and a few other inexpensive components. Perhaps this circuit could best be utilized as a test instrument for automotive mechanics, as most modern American automobiles other than sports cars do not come equipped with this device. On the other hand, the meter could easily be dash mounted for an in-car addition to the instrument panel. The entire circuit can be constructed on a small piece of vector board and mounted directly to the meter terminals. The use of shielded cable to the distributor points is recommended. RF bypassing may be necessary to keep from inducing unwanted whine in the passenger compartment electronics—"Linear Data Manual Volume 2: Industrial." Courtesy of Signetics Company, a division of North American Philips Corporation.

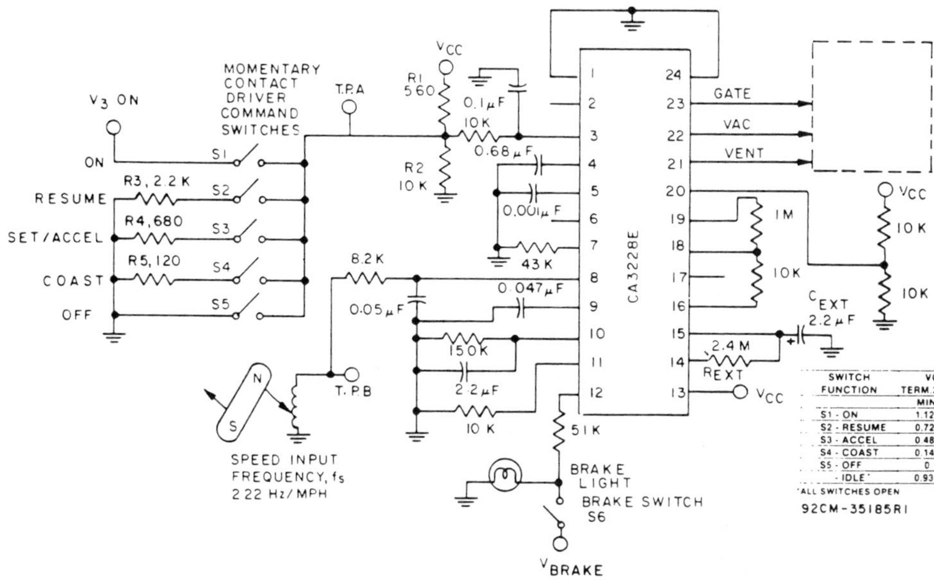

SWITCH FUNCTION	VOLTAGE AT TERM.3 RATIO TO V_{CC}	
	MIN.	MAX
S1 - ON	1.12	—
S2 - RESUME	0.725	0.8
S3 - ACCEL	0.482	0.599
S4 - COAST	0.148	0.272
S5 - OFF	0	0.094
- IDLE*	0.93	0.96

*ALL SWITCHES OPEN

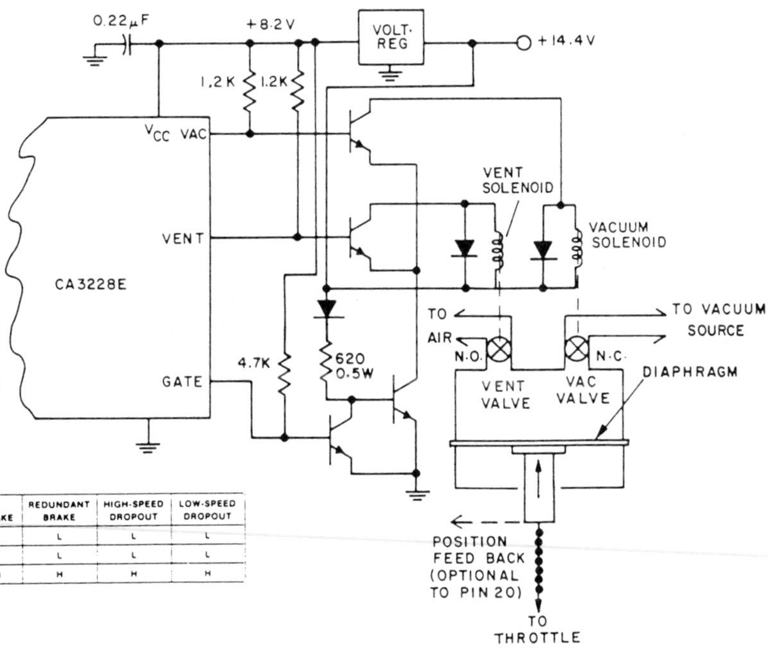

THROTTLE CONTROL

VALVE	SYSTEM MODE		
	ACCEL	CRUISE	COAST
VAC	OPEN	N.C.*	N.C.
VENT	CLOSE	CLOSE*	N.O.

*OPEN OR CLOSED AS REQUIRED TO MAINTAIN SET SPEED ERROR.

CA3228E OUTPUTS

	ACCEL	CRUISE	COAST	BRAKE	REDUNDANT BRAKE	HIGH-SPEED DROPOUT	LOW-SPEED DROPOUT
VAC (PIN 22)	H	L	L	L	L	L	L
VENT (PIN 21)	H	H	L	L	L	L	L
GATE (PIN 23)	L	L	L	H	H	H	H

736 Automobile Speed Control System. This circuit is a complete electronically activated speed control intended for automotive applications. It uses a specialized RCA IC intended specifically for speed control systems. The on-chip system monitors vehicle speed and compares it to a stored reference speed. Any deviation in vehicle speed causes a servo mechanism to open or close the engine throttle as required to eliminate the comparison (speed) error. The reference speed is stored by the driver in a 9-bit counter. The first schematic shows the chip and its associated circuitry. The second shows a suggested mechanical control system to be driven by the speed control circuit.—"RCA Solid State Databook—Integrated Circuits for Linear Applications." Courtesy of Harris Semiconductor.

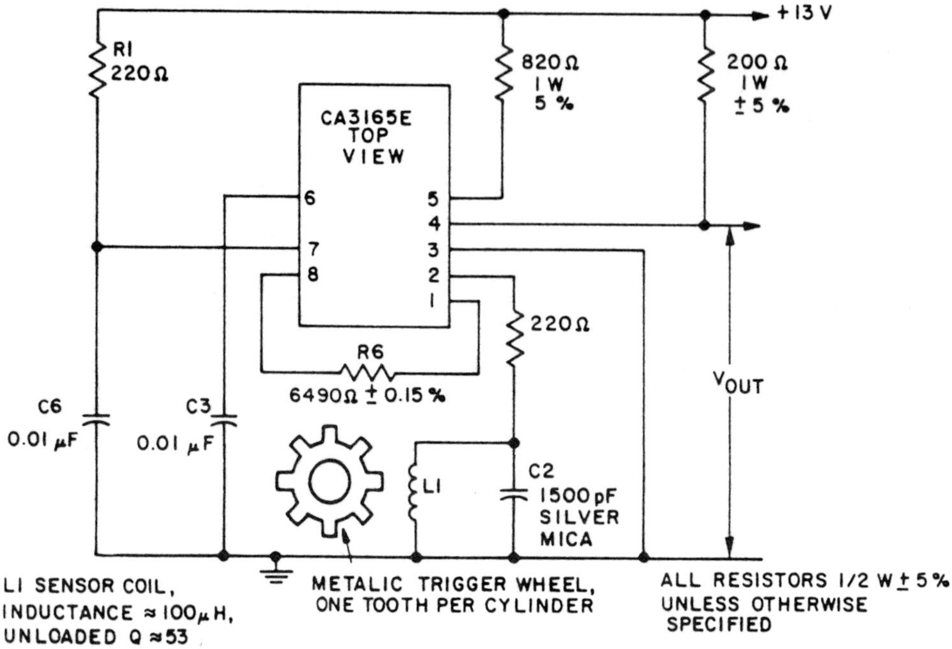

Automotive Ignition System. The RCA CA3165 series is a single-chip electronic switching circuit intended primarily for ignition applications. The oscillator on the chip operates at about 400 kHz (in this application) as determined by the LC circuit composed of L1 and C3. The amplitude of the oscillation is detected on the chip and applied to the Schmitt trigger, which sets the terminal voltage. As a metallic tooth in the rotor passes coil L1, eddy current losses occur which reduce the Q of the resonant circuit and decrease the amplitude of the oscillators to a level that lies below that of a reference in the detector circuit. The output terminals then switch states.—"RCA Solid State Databook—Integrated Circuits for Linear Applications." Courtesy of Harris Semiconductor.

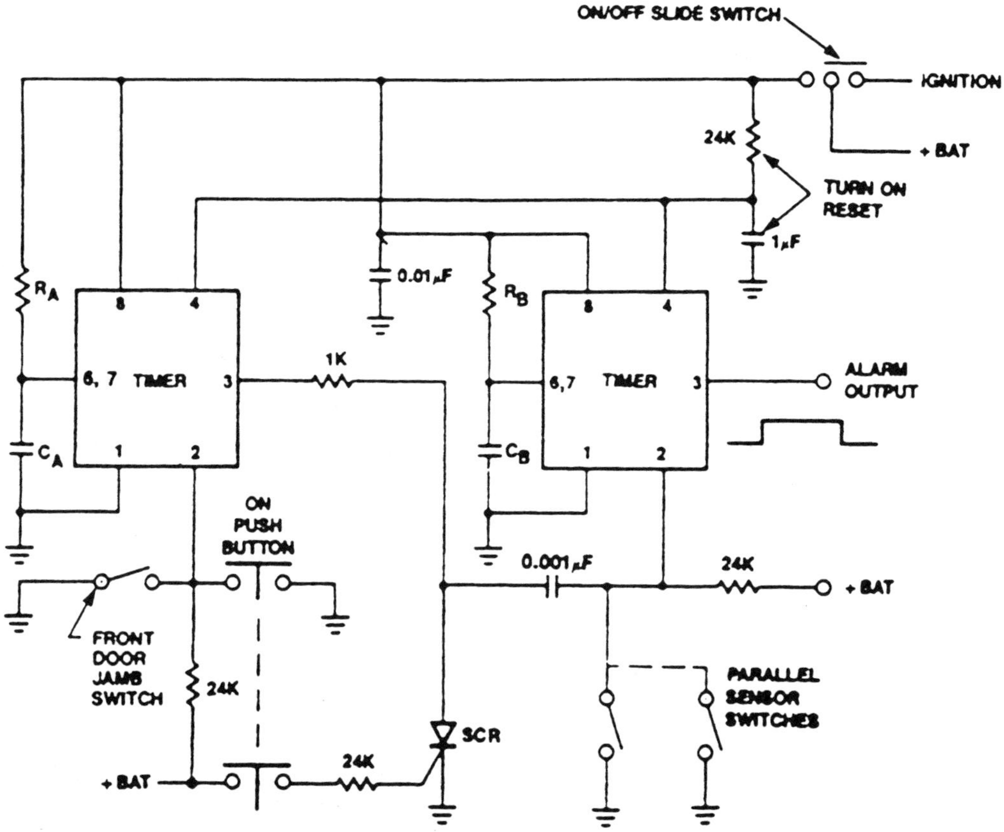

Automobile Burglar Alarm. This alarm circuit uses two Signetics NE555 timer ICs. The first IC (timer A) produces a safeguard delay, allowing the automobile driver to disarm the circuit and eliminating a vulnerable cutoff switch that would be located outside of the automobile. The SCR prevents this first timer from triggering the second timer (timer B) unless the latter timer is triggered by strategically placed sensor switches.—"Linear Data Manual Volume 2: Industrial." Courtesy of Signetics Company, a division of North American Philips Corporation.

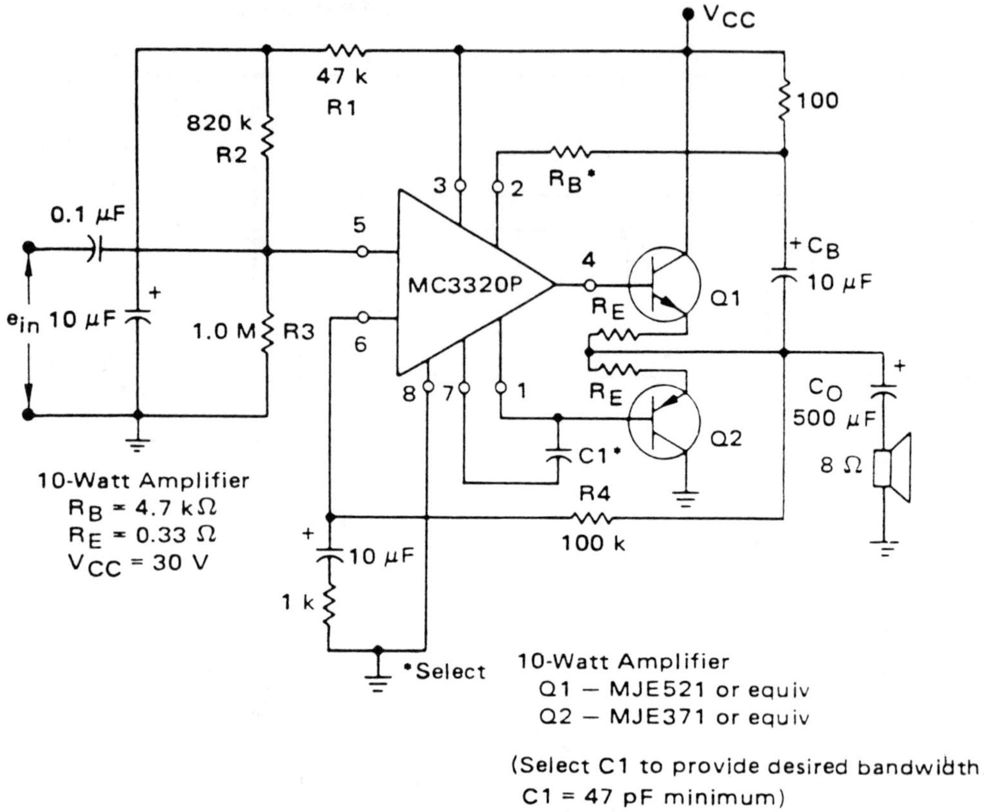

High-Powered Auto Radio Amplifier. Capable of a full 10-W output, the gain of this circuit changes when the value of R4 is varied. The bandwidth is determined by C1. The emitter resistors (R_e) are required at the higher voltages used for 10-W audio amplifiers to provide thermal stability. Do not omit them from the circuit. The actual value of these devices is a function of the heatsink thermal resistance and supply voltage.—"Linear and Interface Integrated Circuits." Copyright of Motorola, Inc. Used by permission.

Chapter 11

Miscellaneous Circuits

All of the circuits that just could not be categorized have been dumped into this chapter. Some are quite obscure and may even be useless for any sensible application. Others may be found to be very useful for that special project for which no known circuit exists. In this chapter you will find tone encoders/decoders, coaxial line drivers, range-detecting circuits, servo amplifiers, and even a music synthesizer.

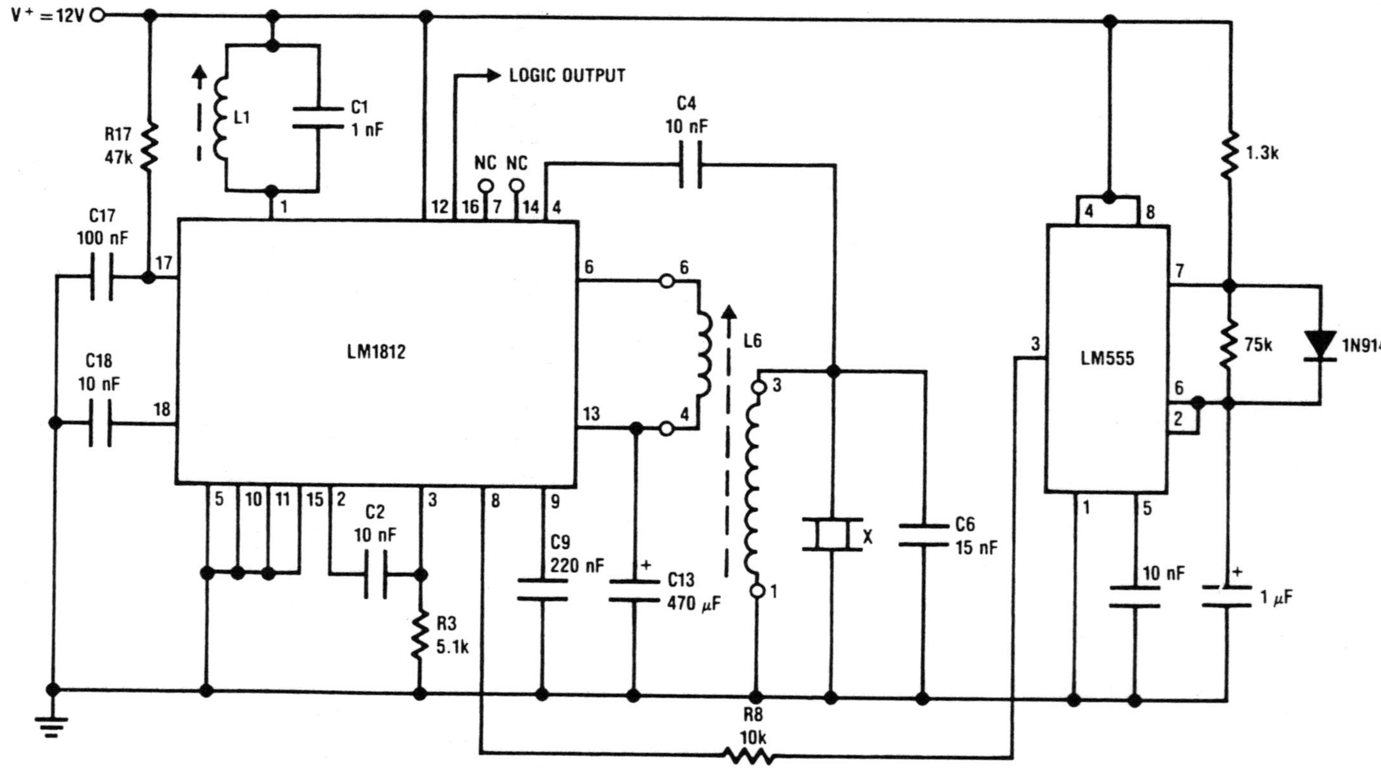

Ultrasonic Ranging System. This configuration operates at 40 kHz and covers a range of 3 to 20 ft. The heart of the circuit is the LM1812 ultrasonic transceiver IC and an LM555 timer. L1 is Toko part #CLN-2A900HM and L6 is part #719VXA-A017AO from the same company. X represents an ultrasonic transducer which is available from Panasonic as part #EFR-OTB40K2.—"Linear Databook III." Reprinted with permission of National Semiconductor.

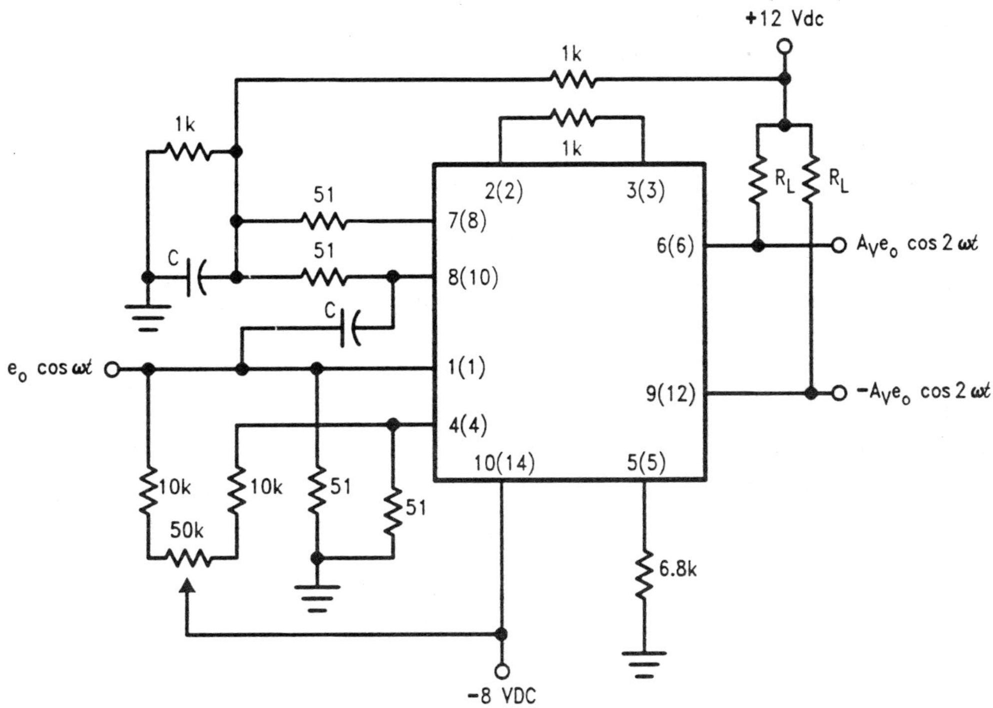

Broadband Frequency Doubler. Using the LM1596 IC, this circuit will double the low-level input signals with very low distortion products. The value of C should be chosen for low reactance at the desired operating frequency. Signal level at the carrier input must be less than 25 mV peak in order to maintain operation in the linear region of the switching differential amplifier. Levels to approximately 50 mV peak may be used with some distortion of the output waveform. If a larger input signal is available, a resistive divider may be used at the carrier input, with full signal to be applied to the signal input.—"Linear Databook III." Reprinted with permission of National Semiconductor.

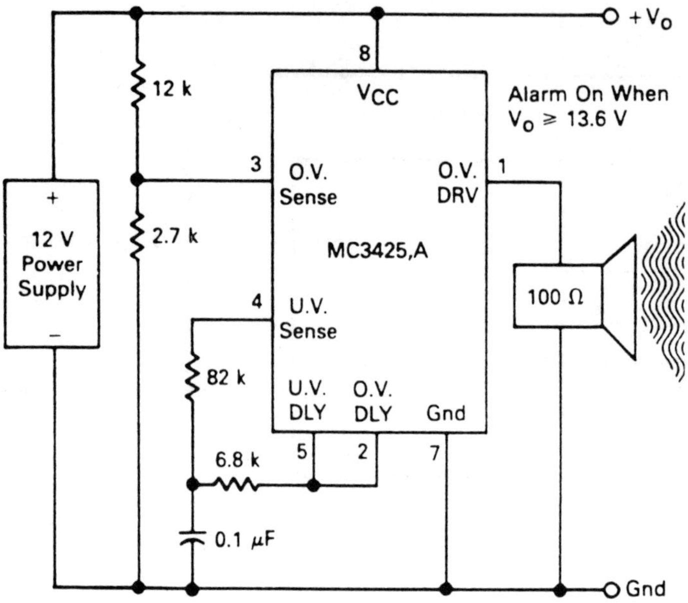

Overvoltage Audio Alarm Circuit. This circuit is designed to monitor a nominal 12-V DC supply and sounds an audio alarm whenever the voltage value rises to 13.6 V DC or greater. This chip is capable of a 300-mA drive output and will operate over a range of 4.5 to 40 V DC, depending on component values. The circuit is straightforward and uses commonly available parts. It drives a 100-Ω sonic transducer which may be mounted along with the circuit in a tiny enclosure.—"Linear and Interface Integrated Circuits." Copyright of Motorola, Inc. Used by permission.

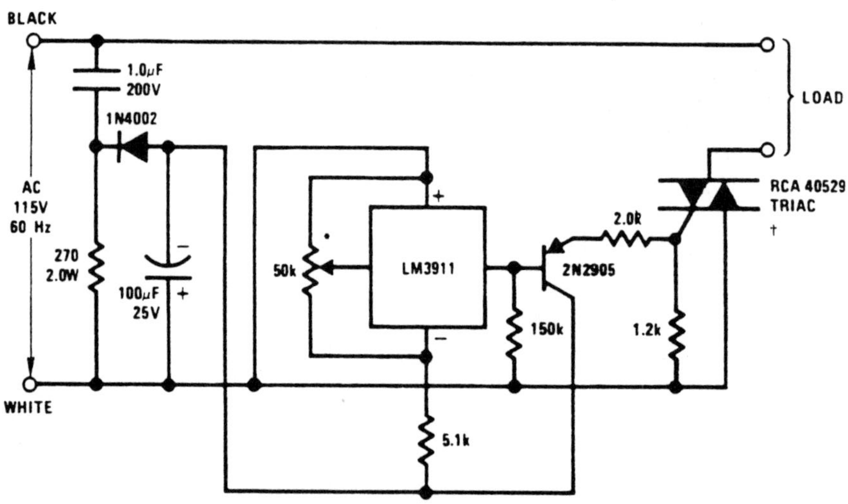

Electronic Thermostat. This triac triggering circuit may be used to electronically activate a fan or air conditioner when the temperature rises above a certain preset level. The LM3911 temperature controller IC is used as the heart of this circuit, and power is received directly from the 120-V AC house current line. The 50-kΩ control is used to set the temperature at which the triac is triggered.—"Linear Databook II." Reprinted with permission of National Semiconductor.

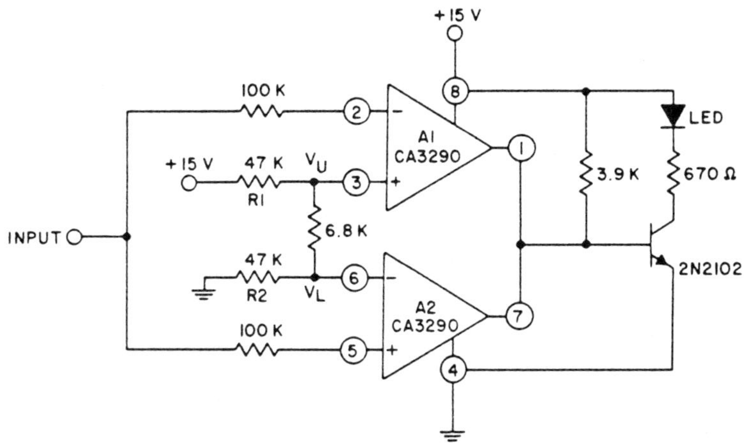

Window Comparator. Both halves of the CA3290 comparator are used in this high input impedance window comparator. The LED will be turned on whenever the input signal is above the lower limit (V_l) but above the upper limit (V_u) as determined by the resistor divider network comprised of R1, R2, and R3. The RCA IC is a BiMOS dual voltage comparator with a MOSFET input and a bipolar output.—"RCA Solid State Databook—Integrated Circuits for Linear Applications." Courtesy of Harris Semiconductor.

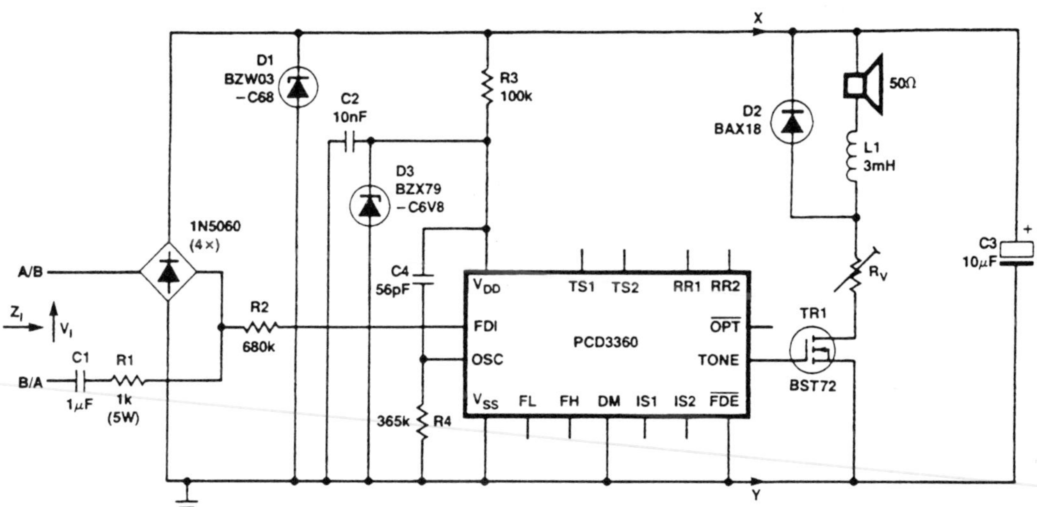

Electronic Telephone Ringer. This interesting circuit uses the Signetics PCD3360 programmable multitone telephone ringer IC in conjunction with a few external components to replace the electromechanical bell in telephone sets. This chip outputs its audio signals to a 50-Ω speaker/audio transducer and connects directly to the telephone line without the need for a transformer. The high operating voltage (line) combined with a class D output stage ensures a high sound level. The diode bridge, zener diode (D1), and resistor R1 protect this ringer circuit against transients up to approximately 5000 V.—"Linear Data Manual Volume 1: Communications." Courtesy of Signetics Company, a division of North American Philips Corporation.

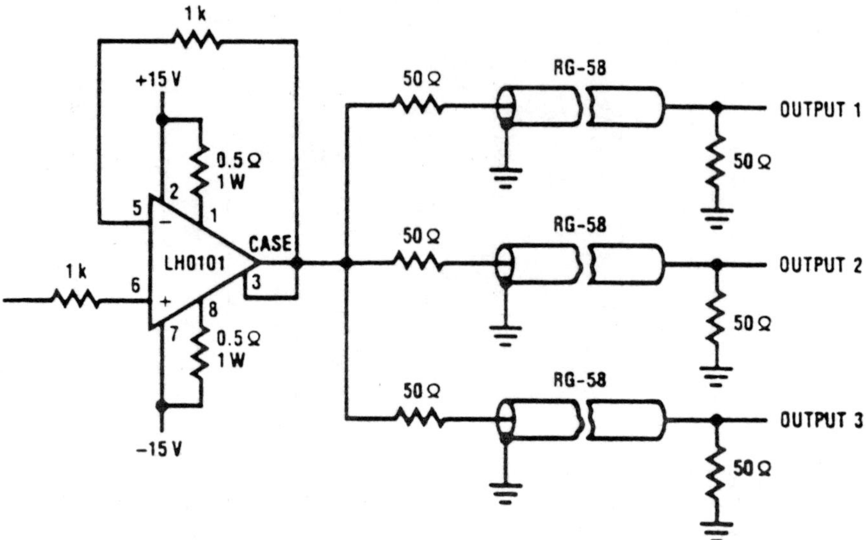

Multiline Coaxial Cable Driver. This circuit is capable of driving up to six coaxial lines. It can drive more if the power op amp is mounted on a heat sink. The LH0101 is a power op amp conservatively rated at 2 A continuous output current. Due to this power capability, this device makes an ideal coaxial cable driver and has adequate bandwidth for most audio and sub-video applications. The high-current, distortion-free output can easily interface a wide range of terminations. Large line capacitance does not present a problem, as there is adequate reserve current capability to charge the capacitance without seriously degrading bandwidth. As shown, the circuit does not include any current-limiting protection against cable shorts. This is a highly desirable feature that should be added.—"Linear Applications Databook." Reprinted with permission of National Semiconductor.

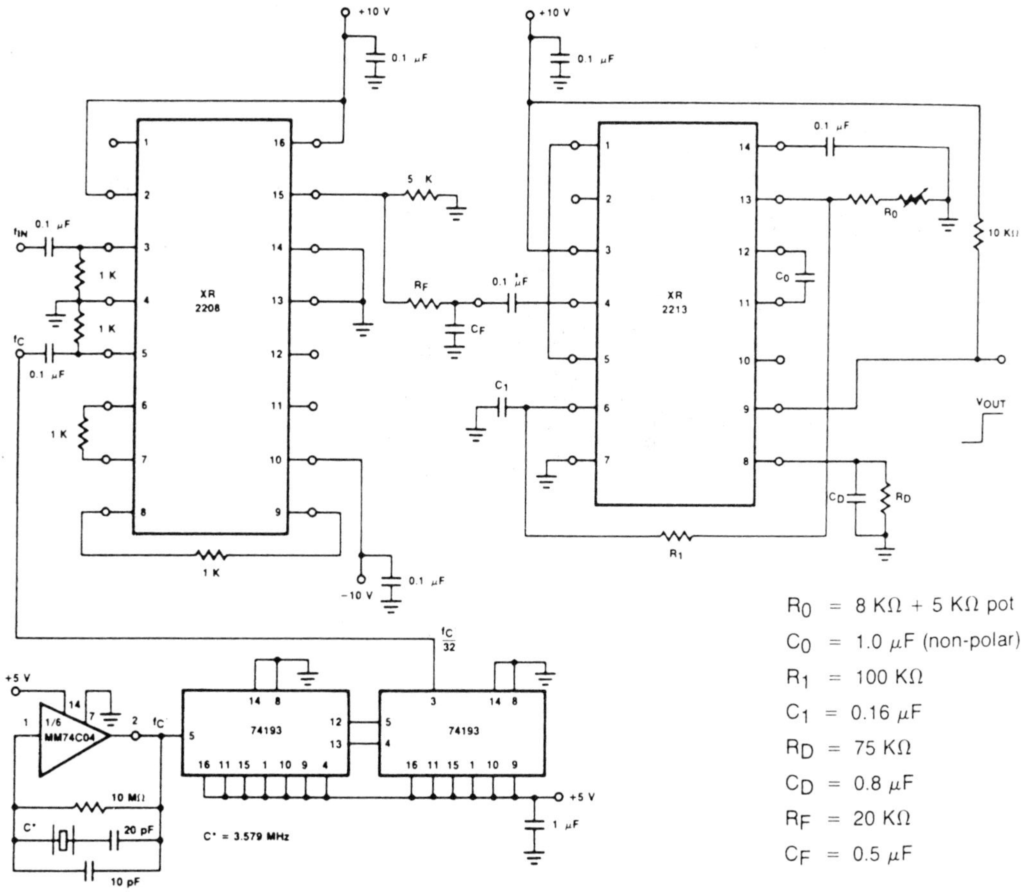

R_O = 8 KΩ + 5 KΩ pot
C_O = 1.0 μF (non-polar)
R_1 = 100 KΩ
C_1 = 0.16 μF
R_D = 75 KΩ
C_D = 0.8 μF
R_F = 20 KΩ
C_F = 0.5 μF

Narrow Band Tone Decoder. Utilizing a PLL in conjunction with a frequency mixer, this circuit is capable of detecting a 1-Hz tone out of a frequency spectrum greater than 1 MHz. It can accept almost any periodic waveform including sine, square, and triangular waves.—"EXAR Databook." Reproduced with the express authorization of EXAR Corporation.

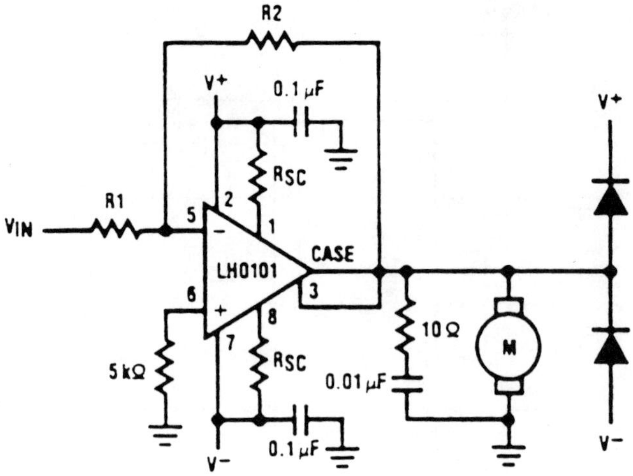

Servo Motor Amplifier. This amplifier will deliver the rated current into the motor, but care must be taken to keep power dissipation within permissible levels. It uses the LH0101 power operational amplifier, which can supply 2 A of continuous output current or 5 A of peak current.—"Linear Applications Databook." Reprinted with permission of National Semiconductor.

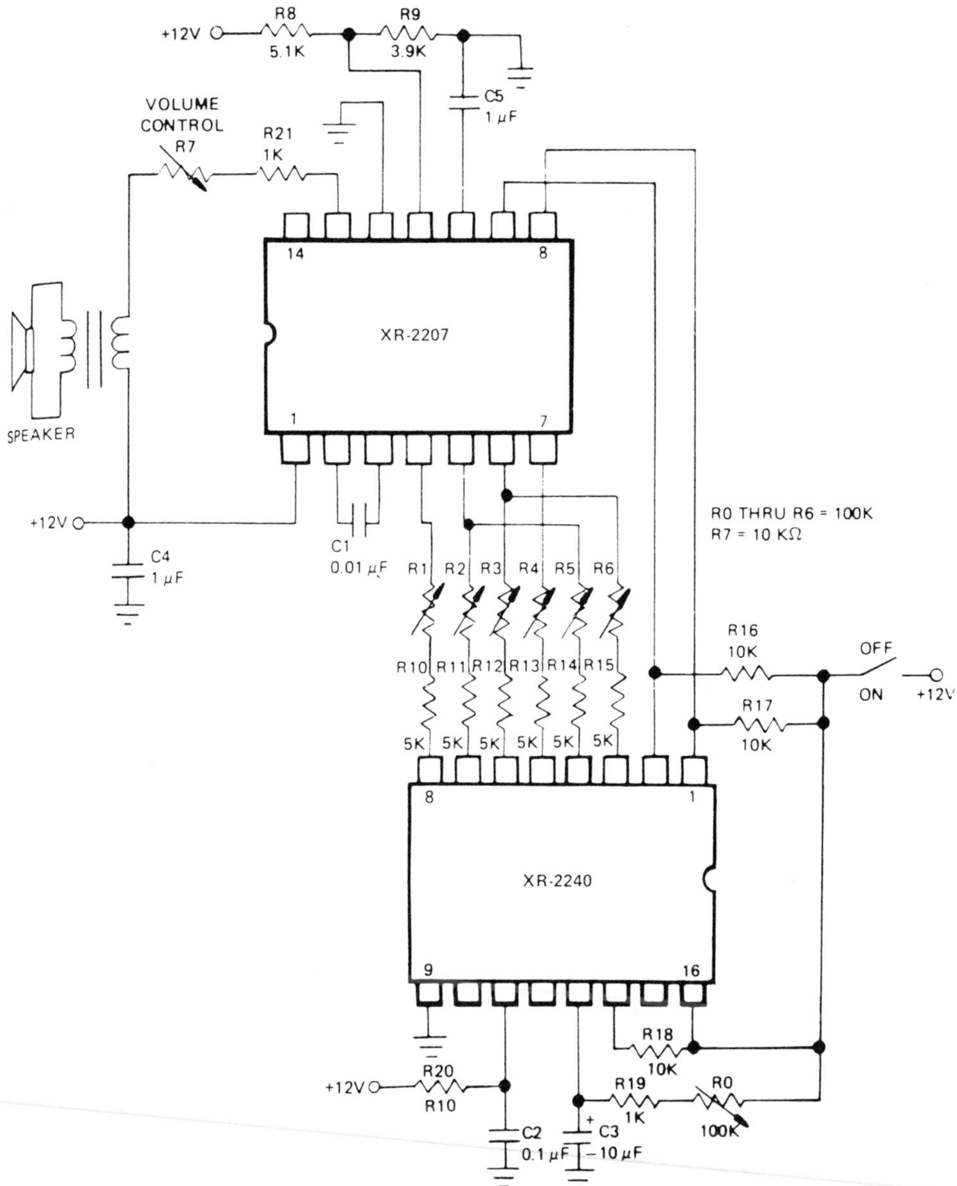

Electronic Music Synthesizer. An EXAR programmable tone generator (XR-2207) is driven by a monolithic counter/timer IC (XR-2240) to produce this minimum-component music synthesizer. All active components are contained within the IC housings, while external passive components are limited to 17 resistors and a handful of fixed capacitors. With this simple circuit, an extremely wide range of audio effects can be generated. It also offers the advantage of being powered by a single-polarity, 12-V DC source.—"EXAR Databook." Reproduced with the express authorization of EXAR Corporation.

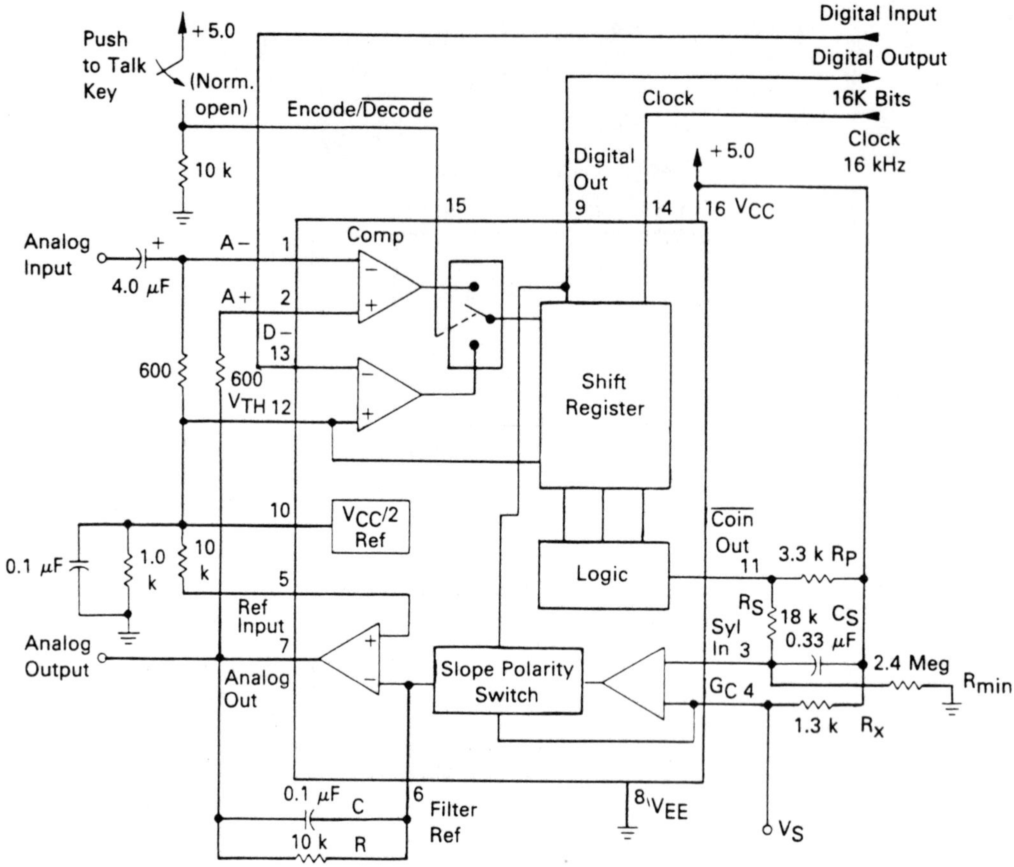

CVSD Voice Encoder. Using the Motorola MC34115 continuously variable slope delta (CVSD) modulator/demodulator IC, this 16-kHz simplex voice encoder is easily constructed. This chip is a general purpose CVSD building block which allows the system designer to tailor the encoder's transmission characteristics to the application. This schematic depicts the most basic circuit possible, most probably rendering it inadequate for high-security applications.—"Linear and Interface Integrated Circuits." Copyright of Motorola, Inc. Used by permission.

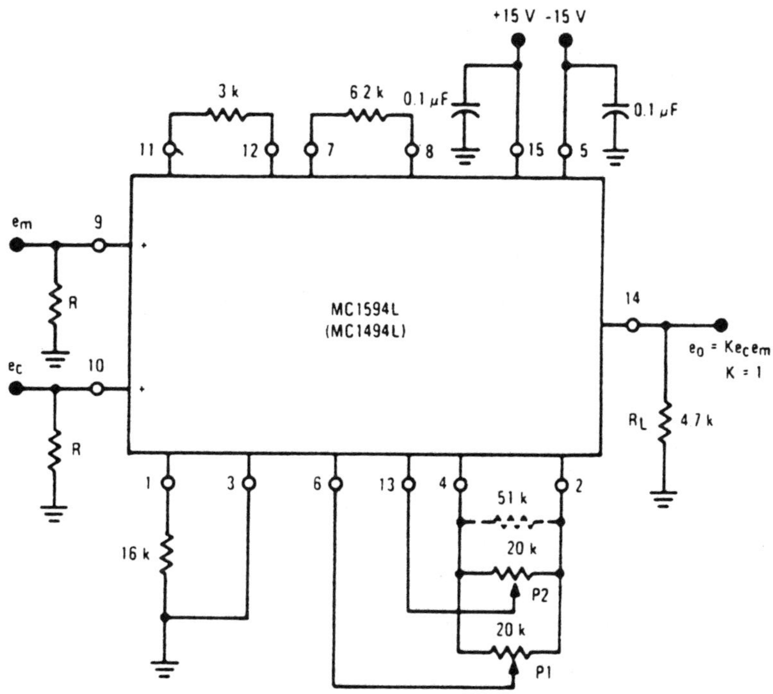

Balanced Modulator. Built around the Motorola MC 1594L four-quadrant multiplier, this circuit produces linear modulation as opposed to other modulation techniques that use nonlinear schemes. Through linear modulation techniques, two sinusoidal inputs will produce an output that contains only two frequencies, the sum and difference of the two inputs. There will be no spectrum centered about the second harmonic of the carrier or, for that matter, any multiple of the carrier. The input resistances are used to supply bias current to the multiplier inputs as well as to provide matching input impedance. The output frequency range is determined by the 4.7-kΩ output impedance and capacitive loading. Assuming a 6-pF load, the small-signal bandwidth is 5.5 MHz.—"Linear and Interface Integrated Circuits." Copyright of Motorola, Inc. Used by permission.

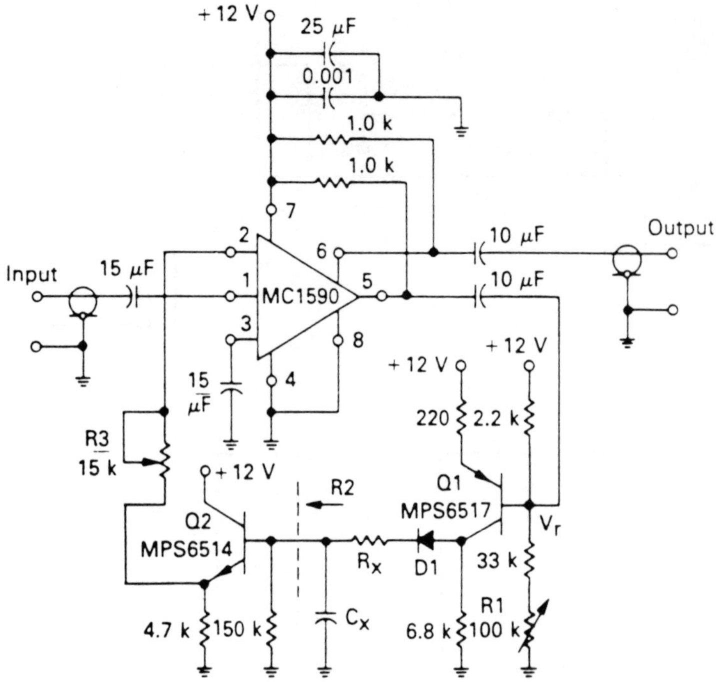

Speech Compressor. This relatively simple speech compressor circuit uses a single IC and can be incorporated in many different types of voice communications systems. The amplifier drives the base of a P-N-P transistor operating in the common-emitter configuration. This portion of the circuit exhibits a voltage gain of approximately 20. R1 varies the quiescent Q point of the transistor so that varying amounts of signal exceed the level V_r. D1 rectifies the positive peaks of Q1's output only when these peaks are greater than $V_r = 7.0$ V. The output is filtered by C_x and R_x. R_x controls the attack time while C_x is involved with both charge and discharge. The emitter follower (Q2) drives the AGC pin 2 of the IC and reduces the gain. R3 controls the slope of signal compression.—"Linear and Interface Integrated Circuits." Copyright of Motorola, Inc. Used by permission.

Circuit Index